The Landforms of
England and Wales

This volume is dedicated to
AMY LOUISE GOUDIE

THE LANDFORMS OF ENGLAND AND WALES

Andrew Goudie

BLACKWELL
Oxford UK & Cambridge USA

First published 1990
First published in paperback 1993

Blackwell Publishers
108 Cowley Road, Oxford, OX4 1JF, UK

238 Main Street, Suite 501
Cambridge, Massachusetts 02142, USA

British Library Cataloguing in Publication Data
A CIP catalogue record for this book is available from the British Library.

Library of Congress Cataloging in Publication Data
Goudie, Andrew,
    The Landforms of England and Wales / Andrew Goudie.
        p.   cm.
    ISBN 0–631–7306–4   ISBN 0–631–16367–0 (pbk)
    1. Geomorphology – England.   2. Geomorphology – Wales.   I. Title.
GB436.G7G69   1990   88–38718
551.4′1′0942 – dc20   CIP

Typeset in 11 on 12 pt Baskerville
by Hope Services (Abingdon) Ltd, Abingdon, Oxon
Printed in Great Britain by
T. J. Press Ltd., Padstow, Cornwall

CONTENTS

PREFACE

Although England and Wales have produced many geomorphologists, these same geomorphologists have not, at least for many decades, produced a general regional text on the geomorphology of England and Wales. Scotland, however, does have such a volume, written by J. B. Sissons, and it is for this reason, in spite of a healthy dose of Scottish genes, that I have restricted myself to a treatment of England and Wales, rather than attempting to cover the whole of Great Britain. Even this, however, may be seen as rather ambitious, not least because of the huge mountain of literature that I have sifted through. What has amazed me in this endeavour is just how few certainties there are. The last decade has seen many hallowed themes eroded (the Calabrian transgression, the Wolstonian glaciation, the concept of tectonic stability, etc.) and yet there are still major gaps in our knowledge. There are, for example, remarkably few descriptive studies, based on quantitative data, of the form of slopes in most parts of the country.

ASG

Acknowledgements

The author and publishers wish to thank the following for permission to use copyright material:

The Arctic Institute of North America for figs 3.2 and 3.3; Edward Arnold for figs 2.8c, 2.11, 6.10 and 8.8; Biulteyn Peryglacjalny for figs 3.4 and 3.5; Blackwell Scientific Publications Ltd for figs 1,11, 2.5, 7.4, 7.5a and 8.14; and with The Institute of British Geographers for fig. 7.4; figs 8.3 and 8.4; with Unwin Hyman for fig. 1.2; G. S. Boulton for fig. 2.13; British Geological Survey for figs 4.4 and 4.6; British Museum (Natural History) for fig. 1.9; Cambria for fig. 2.8; G. H. Dury for figs 4.9 and 6.1; East Midland Geographer for figs 7.7 and 9.4; Elsevier Science Publishers for figs 5.5, 9.6 and 9.7; R. Evans for fig. 9.3; Gebrüder Borntraeger Verlagsbuchhandlung. for figs 6.2, 6.3 and 6.4; The Geographical Association for figs 4.12 and 6.7; Geographical Society of Ireland for figs 8.2a and b; The Institute of British Geographers for figs 5.2c, 5.10, 6.5, 7.5b and 8.13; Institute of Hydrology for fig. 5.7; Longman Group Ltd for figs 4.5, 4.8, 4.10, 4.11, 5.9, 6.6a and b, 9.2 and 9.5; Macmillan Magazines Ltd and B. Buchardt for fig. 1.6; Macmillan Publishers Ltd for fig. 2.12; Manchester University Press for figs 2.8a and 7.3; Nature Conservancy Council and Ordnance Survey for figs 8.6 and 9.9; Thomas Nelson and Sons Ltd for fig. 4.3; The Oleander Press Ltd for fig. 9.10; Oxford University Press for figs 2.2, 2.10, 3.9, 3.11, 3.15 © F. W. Shotton, and 5.8; Royal Geographical Society for fig. 1.8; The Royal Society for fig. 7.8; Society of Antiquaries of London for fig. 8.9; Soil Survey and Land Research Centre for fig. 3.16; E. R. Trueman for fig. 2.6;

Unwin Hyman Ltd for figs 1.12b, 5.3 and 6.9; John Wiley & Sons Ltd for figs 1.3, 3.6, 3.10, 5.4 and 5.6; Yorkshire Geological Society for fig. 2.9.

1

LONG-TERM CHRONOLOGY

Introduction: the growth of denudation chronology

The key to an understanding of long-term landform evolution is denudation chronology and the key to denudation chronology is the recognition of planation surfaces. That such planation surfaces exist in England and Wales is not in doubt, and much of the relief of upland Britain displays even skylines resulting from flattened summits and hillsides developed across the grain of rock structures. The denudation chronologist, however, has to answer three questions about such surfaces (Belbin, 1985): how were they formed? how many of them are there and where are they? when were they formed?

The first attempt to answer the question of how planation surfaces were formed was made in 1846 by A. C. Ramsay, who proposed that the isolated but roughly height-accordant summits of South Wales were a series of relicts cut by wave action. A similar interpretation was made for surfaces in the Peak District by Plant (1866), and for surfaces in the Lake District by Ward (1870).

Ramsay's original diagram (figure 1.1) was explained thus (Ramsay, 1878, p. 497):

> The strata of this area, and, indeed of much of South Wales are exceedingly contorted. The level of the sea is represented by the lower line; and if we take a straight-edge, and place it on the topmost part of the highest hill, and incline it gently seaward, it touches the top of each hill in succession, in the manner shown by the line *bb* . . .

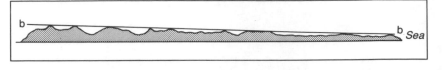

Figure 1.1 A. C. Ramsay's plane of denudation across Wales (from Ramsay, 1846). He believed that the inclined line (*b* to *b*), which touches the tops of hills, represented a former 'great plain of marine denudation'.

> It occurred to me when I first observed this circumstance that, at a period of geological history of unknown date, perhaps older than the beginning of the deposition of the Permian and New Red Sandstones, *this inclined line that touches the hill-tops* must have represented a great plain of marine denudation.

This concept was to have tremendous influence since the development of landscape from an initial surface of low relief was to underlie much British thought on geomorphology for more than a century (Tinkler, 1985, p. 116) for, as late as 1958, Wooldridge (1958, p. 33) was claiming that: 'the chief and best approach to the study of the landforms of an area is by the elucidation of the denudation chronologies, the sequence of distinct and successive phases of land-sculpture preceding the present condition.'

Table 1.1 provides data on the dates, locations and authors of some major studies that were undertaken in England and Wales between 1846 and 1964.

The most extreme manifestation of marine mania is provided by Mackintosh (1869), who culled a whole array of 'evidence' in support of the idea that landforms had been largely moulded by the sea. He looked at the Cotswold scarp near Witcombe and Crickley Hill (Gloucestershire) and saw the embayments as bays, the outliers as islands, and the drift deposits as beaches. Cheddar Gorge and its cliffs were seen as having been moulded by tidal currents, the Brimham Rocks of Yorkshire (called 'insular wrecks') were seen as marine stacks (plate 1), and the Dartmoor Tors were regarded as 'the upper terminations of plains of denudation' formed by the action of sea-waves and currents. He even confused human-made features such as the quarry remnants of the Devil's Chimney on Leckhampton Hill near Cheltenham, and the plough-made terraces (lynchets) of the Chalk Downs, as stacks and

Table 1.1. Selected studies of denudation chronology

Location	Reference
Cornwall	Gullick (1936)
South Wales	Ramsay (1846)
Peak District	Plant (1866)
Lake District	Ward (1870)
Howgill Fells	McConnell (1939a)
Weald	Foster and Topley (1865)
Cumberland	Hollingworth (1935)
SW Wales	Miller (1900)
Forest of Bowland	Moseley (1961)
Staffordshire	Yates (1956)
Ingleborough, Yorks.	Sweeting (1950)
East Devon	Green (1941)
Bodmin Moor	Barrow (1908)
North Cornwall	Balchin (1937)
SW Pennines	Johnson and Rice (1961)
SW Lake District	Parry (1960a)
Middle Trent basin	Clayton (1953)
Northern England	Sissons (1960a)
Exmoor	Balchin (1952)
Wales	Brown (1960b)
Vale of Glamorgan	Driscoll (1958)
E Devon, W Dorset and S Somerset	Waters (1960)
SE England	Wooldridge and Linton (1955)
South Downs	Sparks (1949)
Hampshire Basin	Everard (1957)
River Exe	Kidson (1962)
Alston Block	Trotter (1929)
SW England	Brunsden et al. (1964)
Yorkshire Wolds	Lewin (1969)

beaches respectively. It is perhaps not surprising that his views met waves of resistance in the pages of the *Geological Magazine* during the late 1860s.

However, by the turn of the century an alternative explanation for planation surfaces became available, for the Davisian peneplain concept was based on the ability of long-continued subaerial denudation to cause slope decline through time. Davis developed his ideas in the eastern USA, but applied them to Britain a few years later (Davis, 1895). He was especially interested in the arrangement of the English rivers which (p. 128):

. . . present certain geographical features of systematic arrangement that seem capable of explanation, because their arrangement falls into so close accord with the expectations concerning river growth that are deducible from the geological history of the region. In short, the rivers of today, in the mature stage of the present cycle of denudation, appear to be the revived and matured successors of a well-adjusted system of consequent and subsequent drainage inherited from an earlier and far-advanced cycle of denudation.

Plate 1. The Brimham Rocks, Yorkshire, were regarded as ancient stacks eroded by the ocean. (Author)

He went on (p. 139):

> The uplands of the Oolite and the Chalk seem to me to be
> remnants of a peneplain of subaerial denudation, for the
> reason that their drainage is accomplished in great part
> by subsequent streams . . . as should be the case if the
> present streams are the revived successors of those of a
> former cycle of atmospheric denudation; and not by
> superposed streams imperfectly adjusted to the structure,
> as should be the case if the region had been denuded by
> marine action, and then elevated to its present altitude
> above sea-level.

Following Davis's intervention it became important to identify and
differentiate platforms from peneplains and the following criteria
were used (see, for example, Brown, 1960b, pp. 22–4):

1 Patches of marine deposits resting *in situ* upon several expanses of
  flat are *a priori* evidence of the wave-cut origin of the erosion
  surface to which those flats belong.
2 The presence of a degraded cliff at the rear of the flat, traceable
  at a constant height from flat to flat is suggestive of a marine
  origin.
3 A peneplain may possess greater surface irregularity than a
  platform.
4 The difference between a monadnock rising above a peneplain
  and an island rising above a platform is that an island will
  possess a degraded cliff at a constant height whereas a
  monadnock will not.
5 On a peneplain, the drainage should become well adapted to
  geological struture whereas on a marine platform the reverse will
  be the case and streams transverse to the structure should be
  much in evidence.

Hollingworth (1938) undertook a statistical analysis of altitudinal
data from topographic sheets for large areas of highland Britain
and came up with the following conclusions (pp. 75–6):

> The principal levels of planation, at 430 feet, 730–800 feet
> (possibly two distinct shelves), 1000–1070 feet, and 2000
> feet, occur throughout the western and northern parts of
> Britain. From their correspondence in altitude and in
> relative size and position, and from their relation to

subsidiary shelves at intermediate levels, it is deduced that the platforms at any one level are contemporaneous and represent an unwarped shelf or notch formed during a halt in the emergence of the land . . . It is difficult to envisage a subaerial origin of so widespread a series of platforms of small breadth and limited vertical range, and their formation as coastal marine-cut platforms seems more probable . . . The absence of warping over so large an area is more readily accounted for by the hypothesis of a stable land and falling sea-level than by that of uniform uplift of a land mass.

Classical denudation chronology reached its peak in the work of Wooldridge and Linton (1939, 1955) in south-east England, and of Brown (1960b) in Wales. The complexity of this work is well illustrated by the conclusions to Brown's study (pp. 169–70), in which 18 stages of evolution are outlined:

1 Armorican orogeny.
2 Permo-Trias period of erosion accompanied by:
3 Deposition of the Permo-Trias in the basins around Wales. The main outline of the Welsh block roughed out.
4 Peneplanation (probably more than once), accompanied by transgressions of Jurassic and Cretaceous seas.
5 Frequent interruptions of this process when regressions occurred, as during the deposition of the Inferior Oolite and at the close of Jurassic times.
6 The submergence and burial of Wales beneath Mesozoic deposits completed by the Cenomanian transgression.
7 Early Tertiary (pre-Eocene) uplift. Wales formed the eastern and southern flanks of a dome whose apex was in Snowdonia. It is probable that some relief was afforded by movements along marginal faults at Malvern, in the vale of Clwyd, and between Cardigan Bay and the Welsh uplands. Dykes were probably intruded in north Wales.
8 The initiation of the drainage upon a cover of Cretaceous rocks and the early Tertiary cycle of erosion.
9 The stripping back of the Mesozoic cover to reveal the wave-trimmed Mesozoic peneplain. The superimposition of the radial drainage upon the Palaeozoic rocks beneath. The end product of the cycle was the 'summit plain', a sub-Mesozoic surface exhumed and further eroded in early Tertiary times.

10 Middle Tertiary earth movements. 'En echelon' type folding in the vale of Glamorgan. Renewed uplift, probably decreasing in amount northwards.

11 The High Plateau cycle was initiated by this middle Tertiary uplift and ran its course in Miocene times. The end product was the High Plateau.

12 The formation of the Middle Peneplain. The negative change in base-level from the High Plateau was probably a eustatic one.

13 The formation of the Low Peneplain following a further eustatic drop in base level. The formation of the three peneplains was accompanied by a progressive adaptation of the drainage to the structures in the Palaeozoic rocks of Wales.

14 The transgression of an early Pleistocene sea to a height of just below 700 feet and the cutting of the '600-foot' platform.

15 Regression of the early Pleistocene sea and the initiation of new stream courses across the emergent platform.

16, 17, 18 The cutting of the '400-foot', '300-foot', and '200-foot' platforms in the Pleistocene, perhaps during interglacial periods. The successive falls in base-level from the 600-foot platform caused waves of back-cutting to be transmitted up the rivers, and river capture was more common at this than at any previous time.

In the last decade or so, another possible mechanism for planation surface development in the Tertiary has emerged. This is extreme tropical planation, leading to the formation of inselberg-studded etchplains. Such planation is achieved by a combination of deep chemical decomposition and effective surface wash under tropical conditions with a long rainy season (6–9 months). The prime champion of this mechanism was the German geomorphologist, J. Büdel, who maintained that the Pleistocene had only 'lightly overprinted' the basic relief shaped under tropical conditions: 'In the Early Tertiary this relief type extended even into the polar areas, and its traces can still be found in all climatic zones. For due to their flatness, relict planation surfaces at all altitudes are the most durable of all terrestrial relief forms' (Büdel, 1982, p. 253).

Relict deep weathering phenomena do occur in the British Isles and duricrusts of possible tropical or subtropical type occur in association with Tertiary beds. These include the silcretes of southern England (Summerfield and Goudie, 1980) and the well-developed laterites and bauxites of the Antrim Plateau in Northern Ireland (Smith and McAlister, 1987). In North Wales and

Derbyshire the Mio-Pliocene Brassington formation consists of pockets of what appear to be products of tropical weathering (Brown, 1979, p. 460), while in Cornwall Walsh et al. (1987) believe they have identified pockets of Miocene saprolite on an etchplain of tropical and subtropical origins. Similarly, Battiau-Queney (1984) claims to identify inselbergs, tropical planation surfaces and tropical weathering products in Wales. In East Devon Isaac (1983) has attributed the 'Plateau Deposits' to lateritic deep weathering in the early Tertiary. Kaolinitic ball clays are widespread in south-west England, and these too are thought to form most effectively under hot, wet conditions.

Developing these tropical ideas, Battiau-Queney (1984) has criticized the model for Wales proposed by Brown (1960b). In its place she proposes that, since it emerged in mid-Cretaceous times, the Welsh Massif has suffered a long subaerial evolution, with phases of tropical climate favouring powerful chemical weathering and inselberg development. In the Neogene, vertical movements reactivated ancient structural weaknesses, creating the present Welsh landscape. Caledonian or pre-Caledonian tectonic patterns (see p. 150) controlled vertical movements associated with the onset of spreading in the Rockall Trough and the creation of the North Atlantic Ocean. The Welsh Massif was divided into major structural units, each block acting independently with its own rate of uplift and denudation. Such a view also contrasts vividly with Hollingworth's view (see p. 9) of tectonic stability.

Pre-Tertiary landforms

Pre-Tertiary landforms may be important in some areas of the world – for example on some of the resistant quartzitic surfaces in Australia – but in Britain this is not the case. Indeed Brown (1979, p. 460) has gone so far as to state: 'It is highly unlikely that any facet of the present surface of Britain has been inherited from pre-Tertiary times. Nothing as old as 63,000,000 years could possibly have survived intact without burial and exhumation; even surfaces having little or no slope would at least have suffered substantial lowering by chemical denudation'.

The caveat that some landforms might be present in the landscape today as a result of a history of 'burial and exhumation', is, however, an important one. This is particularly true with respect to the existence of possible exhumed Triassic landscapes of arid zone type. The classic location for landforms of this genre, though by no means the only one, is Charnwood forest in the East Midlands, as was noted by Watts (1910, p. 783):

> The steep craggy slopes of little weathered rock, partly buried in Breccia, would agree with the type of landscape we might expect to find in a dry desert region such as this is likely to have been in Triassic times . . . It appears to the writer that many of the features of the upland and old rock landscapes of Charnwood have emerged practically uninjured on the washing away of the (Keuper) Marl, and that the action of denudation has been chiefly to *develop the latent image* of the Triassic landscape hidden in the Marl, and not to destroy it on emergence.

Another area where exhumation has been postulated as a cause of exposure of Triassic landscapes is Wales, where Jones (1931) postulated that the widespread upland plateau surfaces of Wales may represent an ancient Triassic peneplain which owes its remarkable state of preservation both to the great resistance of the underlying Palaezoic rocks and to its subsequent protection beneath Mesozoic strata.

There is also good evidence for massive denudation of the rocks of the south-west peninsula in the Permian and Triassic. This is provided by the late Carboniferous to early Permian Dartmoor granites which must originally have been intruded at considerable depths below the surface – probably 5–9 km. Debris from these granites occurs in early Permian sediments, by which time erosion must have cut down to the top of the pluton. Material from this source is also present in Triassic conglomerates and subaerial erosion may have continued into the Jurassic and Lower Cretaceous (Green, 1985).

The Cretaceous transgression

In the Upper Cretaceous (*c.*70–100 million years ago), the Chalk was deposited. It is an extremely pure marine limestone that accumulated to a thickness of up to 550 m in at least 210 m of water (Jones, 1985, p. 25). It is probable that when it was deposited Britain, with the exception of a few islands in high-altitude locations, suffered almost total submergence. The Cretaceous submergence probably caused severe marine planation of underlying structures before the Chalk was deposited, and the subsequent deposition of a thick cover of Chalk on top of this marine unconformity effectively sealed off the past. It is for this reason that it is crucial to study the Tertiary if one is to understand the landscape of England and Wales.

Relicts of this almost ubiquitous blanket of Chalk occur as remnant outliers in Ireland and parts of the West Country (for example, between Sidmouth and Lyme Regis), and as major exposures in southern and eastern England.

There is, however, some debate as to how extensive the Chalk cover was, particularly in Wales. Battiau-Queney (1984) has posed the question that if Chalk was extensive over Wales why is it that no residual deposits (especially flints) have been found, either on the surface or, particularly, in karstic pockets or pipes?

Some of the major stages in the palaeogeographic evolution of the British Isles in the Tertiary are shown in figure 1.2. Dates of major events are given in table 1.2.

When the Chalk sea retreated as a result of tectonic uplift, a new-land area became exposed. On this a consequent drainage system developed comprising many large rivers, which drained primarily

Table 1.2 The Tertiary's divisions

Date (million years ago)		Division	Relict
2		Pliocene	Coralline crag, Lenham beds, St Erth beds
	NEOGENE	↓	
7		Miocene	
		↓	
26			
		Oligocene	Hamstead beds
			Bembridge and Osborne beds
	PALAEOGENE	↓	Upper and Middle Headon beds
37		Eocene	Lower Headon beds
			Barton beds
			Bagshot beds and Bracklesham beds
		↓	London clay
53		Palaeocene	Woolwich and Reading beds
65		↓	Thanet Sands

towards the present North Sea (figure 1.3). The Chalk cover has subsequently been stripped from large areas but the drainage pattern developed on it has probably been superimposed on to the underlying Mesozoic and Palaeozoic rocks. As Small (1970, pp. 258–9) has explained:

> Subsequently, adjustment to the rock outcrops and structures thus exposed has reached an advanced stage. Even so, many instances of streams which disregard structural trends, and which may therefore be attributed to superimposition from the Chalk, may be identified in areas to the west and north of the present Chalk margins. In Wales, many important rivers contain long stretches in which they run broadly from north-west to south-east, despite the fact that the Caledonian and Hercynian structures are aligned from north-east to south-west (central Wales synclinorium and Teifi anticlinorium) or from east to west (the south Wales coalfield synclinorium).

This concept of drainage superimposition from a Chalk cover has a long history, and was most forcefully championed by Linton who believed that two main consequent systems drained much of Wales and Midland England: a proto-Trent (which included the upper stretches of the Welsh Dee) and a proto-Thames. Linton (1951, p. 454) wrote:

> The evidence of superposition of the drainage from a vanished cover on to an undermass of unrelated structure is so widespread and consistent . . . that it seems impossible not to conclude that the drainage of the whole region from South Wales to The Pennines, and perhaps to central Scotland, originated by superposition from a single widespread cover formation at a single epoch. If this conclusion be accepted there is only one rock formation which possibly could have played the role of cover rock, namely the Chalk.

In their classic study, Wooldridge and Linton (1955, p. 1) were convinced of the importance of the Tertiary:

> The shaping of the British landscape has been the work of Tertiary times, a period now indicated, by radioactive

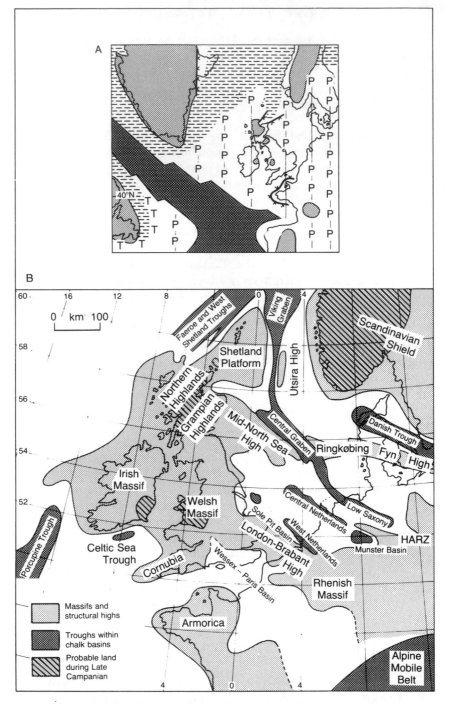

Figure 1.2 The palaeogeography of the British Isles and environs at various stages from the Cretaceous to the Miocene (modified after Anderton et al., 1979, figures 16.4., 16.8. and 16.10 and Hancock, 1986, figure 7.7.): (A) generalized palaeogeography in the Upper Cretaceous (Campanian); (B) detailed Upper Cretaceous geological setting; (C) Palaeocene palaeogeography. Note the position of the

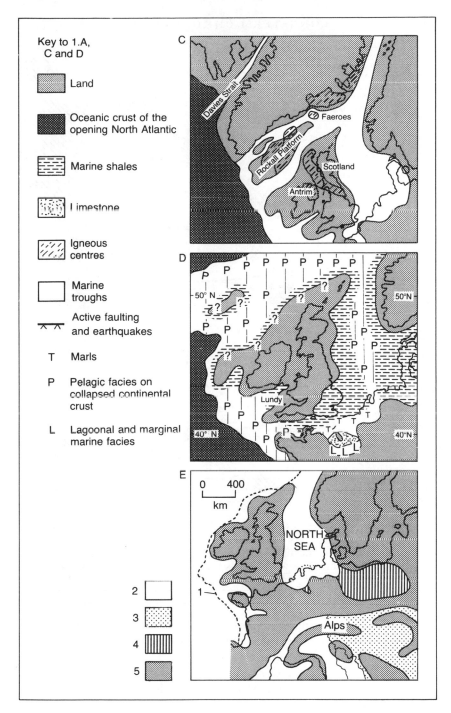

British Isles with respect to Greenland and the widespread nature of igneous activity; (D) Eocene palaeogeography, key as for (A); (E) Miocene palaeogeography, 1 = present-day 200 m isobath, 2 = sea, 3 = the Para-Tethys sea, 4 = basins of continental sedimentation, 5 = land.

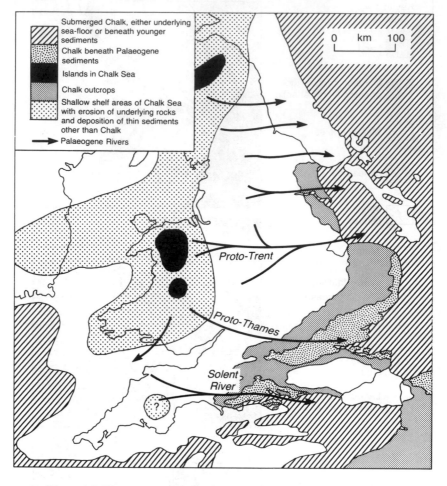

Figure 1.3 The present distribution of Chalk, the palaeogeography of
the Upper Cretaceous (Chalk) sea, and the presumed courses of the
Palaeogene consequent drainage pattern that developed on the newly
emerged Chalk. Areas where Chalk is thought to have been deposited
but subsequently removed by denudation are shown in white
(modified after Jones, 1985, figure 1.15).

dating, as having a duration of some 65–70 millions of
years. In this long interval the mountains were shaped,
the valleys carved, and the coastlines developed to their
present form. It is in this sense that Tertiary time in Britain
means more for geomorphology than for stratigraphical
geology. The story of sedimentation is scant, but the

interval of time represented is vast, vast enough for the evolution of the whole British landscape as we know it.

They were also convinced of the importance of south-east England for elucidating this story, because of its Tertiary deposits and record of erosion: 'Our region thus serves as a comparator for the other regions of Britain and in high degree can claim the role of type area for British geomorphology' (Wooldridge and Linton, 1955, p. 1).

The basic features of their model were three major phases of planation producing three major planation surfaces (figures 1.4 and 1.5): a sub-Eocene surface (now exhumed), a subaerially produced Mio-Pliocene peneplain and a marine plane of Plio-Pleistocene age. Much subsequent work has sought either to identify these three surfaces or to refute their existence.

The early Tertiary (Palaeogene)

Much of the British Isles became land during the Palaeogene and the Chalk blanket became eroded. The unresistant Chalk probably suffered rapid attack and was removed over large areas at a relatively early date. Denudation was then able to attack under-lying lithologies, and would have been able to contribute debris to the thick sedimentary sequences that accumulated in the North Sea Basin and in the London and Hampshire Basins. They also provided terrigeneous clastic sediments to the fault – controlled basins that developed in the west (e.g. Bovey Tracey, Petrockstow, the Celtic Sea, the British Channel and Cardigan Bay).

Indeed, in south-west England the Bovey Formation provides useful evidence for reconstructing the Palaeogene geomorphological environment (Green, 1985). It contains kaolinitic clays which are derived from the weathering of granite and Palaeozoic sediments, indicating that the basement rocks were quite extensively exposed in Eocene times. Likewise, although Chalk-derived flint is a major component of the Upper Palaeocene Reading Beds, their clay mineralogy and pebble content indicate that the basement rocks of south-west England were exposed in Reading Beds times. Moreover, the uplands of east Devon, south Somerset and west Dorset possess a cover of Tertiary residual deposits and Tertiary sediments, including lateritic weathering products and silcrete consistent with

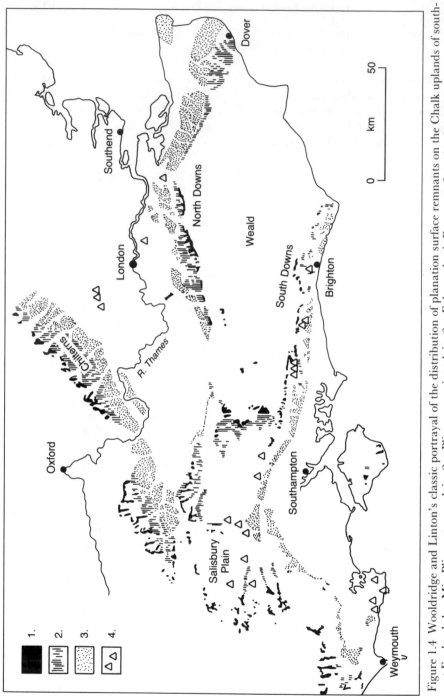

Figure 1.4 Wooldridge and Linton's classic portrayal of the distribution of planation surface remnants on the Chalk uplands of south-east England: 1 = Mio-Pliocene peneplain; 2 = Pliocene marine plain; 3 = Exhumed sub-Eocene surface; 4 = Summits rising to the level of the marine plain (modified from Wooldridge and Linton, 1955, figure 17).

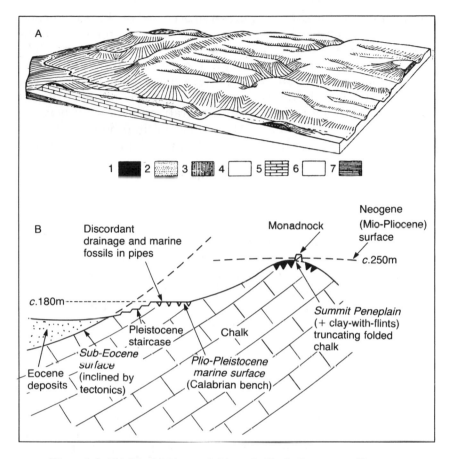

Figure 1.5 (A) Wooldridge and Linton's block diagram to illustrate 'The three facets of a fully developed chalk cuesta: Mio-Pliocene land surface; Pliocene marine bench; and exhumed sub-Eocene plane'. The approximate dimensions of the block are 11 × 26 km. 1 = residual clay-with-flints; 2 = Pliocene marine deposits; 3 = Eocene clay and sand; 4 = Upper Chalk; 5 = Middle Chalk; 6 = Lower Chalk; 7 = Gault clay (after Wooldridge and Linton, 1955, figure 16). (B) Schematic diagram of the classic Wooldridge and Linton model.

the idea 'that the denudation of the Chalk was largely affected under tropical or subtropical conditions before the end of the Palaeocene' (Green, 1985, p. 69).

The deposition that occurred in south-east England records the interplay between marine transgressions from the North Sea and the subsequent progradation of a terrigenous coastal plain. The most widespread marine transgression occurred during the Eocene, and the competitive interplay between marine and non-marine environments continued until the end of Eocene times. The Oligocene sequence, preserved on the Isle of Wight, illustrates the continuation of transgressive/regressive cycles. However, the marine influence decreased with time. Indeed, in the Miocene uplift it produced a major gap in the stratigraphic record between the youngest Oligocene in the Hampshire Basin (*c.*30 million years ago) and the Pliocene beds of East Anglia (*c.*3 million years ago) (Anderton et al., 1979, ch. 16).

The importance of the Palaeogene has been summarized by Jones (1985, pp. 33–4):

> The various lines of evidence for landscape evolution during the Palaeogene suggest that by the mid-Oligocene (30 Ma), the general form of Britain had been largely blocked out, much of the tectonic deformation had been completed, and denudation had removed much of the chalk cover and had already achieved considerable sculpturing of the underlying strata in upland areas.

The sub-Palaeogene surface

One of the most important Tertiary erosion surfaces, particularly in south-east England, is a largely marine-trimmed surface upon which deposition of Eocene sediments (including the London Clay) occurred. This surface, which Wooldridge and Linton (1939, 1955) called the 'sub-Eocene surface', and which they believed had been deformed by post-depositional tectonism, is now generally termed the 'sub-Palaeogene surface' (Jones, 1981, p. 72):

> It is clear from the differences in age of the Palaeogene deposits found resting directly on the Chalk that the sub-Tertiary unconformity is diachronous and that the continued use of the term 'sub-Eocene surface' is both incorrect and undesirable. In the London and Hampshire–Dieppe Basins the basal surface is overlain by Palaeocene deposits and the exhumed surface exposed adjacent to these outcrops is also of Palaeocene age. It is only on the higher flanks of the Hampshire Basin that there exists

clear stratigraphic evidence for a basal surface of Eocene age. As a consequence, the term 'sub-Palaeogene surface' should be used to describe this unconformity.

An alternative name is 'early-Tertiary surface' (Small, 1970, p. 276.).

The nature of early Tertiary climates

Given the possibility that the planation of Britain was achieved under relatively warm climatic conditions, the question arises: what independent evidence is there for the nature of early Tertiary climates in Britain and Western Europe?

The prime evidence is provided by plant remains in Tertiary sediments. The classic interpretation was that of Reid and Chandler (1933), who believed that the Palaeocene London Clay flora compared most suitably with the present-day vegetation of the Indo-Malayan region, indicating a tropical rainforest environment during early Eocene times in southern Britain, with a mean annual temperature that may have been in excess of 21°C. However, as Daley (1972, p. 187) has pointed out, 'Such a climate, with its markedly uniform character throughout the year, could not have existed at the latitude of southern Britain during the Eocene (40°N), since seasonal climatic variations would undoubtedly have occurred.' In addition there is a small proportion of extratropical types within the London Clay flora. One possibility is that the mixture of tropical and extratropical plants may have resulted from a type of climate not represented at the present day. Daley envisages that given frost-free conditions and high rainfall levels tropical plants may have become established in low-lying damp areas even though temperatures were not truly tropical.

Nevertheless the picture of early and middle Eocene warmth is confirmed by more recent studies. For example, Collinson and Hooker (1987) report that beds of that age contain the highest proportion of potential tropical flora (up to 82 per cent) and the smallest proportion of potential temperate nearest living relatives (up to 42 per cent). Floras from the late Palaeocene and latest Eocene and early Oligocene contain a smaller proportion of potential tropical elements (e.g. 70–82 per cent). Collinson and Hooker (1987, p. 267) consider that: 'A great many of the near living relatives of the fossils co-exist today in paratropical rain forest in eastern Asia. Many are large forest trees, others lianas. The greatest diversity of these elements in the Early and Middle Eocene suggests the presence of a dense forest vegetation'. Tree rings were also particularly large, indicating rapid growth in response to high temperatures (Creber and Chaloner, 1985).

Confirmation of the existence of Eocene warmth comes from the isotopic evidence obtained from cores in the North Sea and elsewhere (Buchardt, 1978). Figure 1.6 indicates the estimated palaeotemperatures for the whole Tertiary in the North Sea basin and suggests that not only was the Eocene exceptionally warm, but that there was a severe climatic deterioration at the end of the Eocene. Before that deterioration, temperatures in Britain may have been on average 20°C higher than they are today.

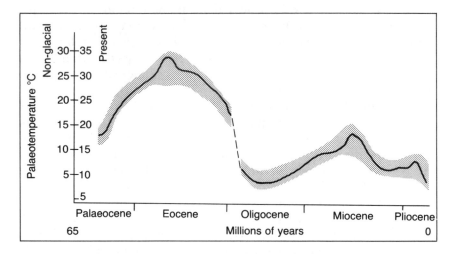

Figure 1.6 The climatic history of the Tertiary, including the post-Eocene climatic decline, as calculated from oxygen isotope values of shells in the North Sea (after Buchardt, 1978). The difference between the non-glacial and the present temperature scales is caused by the shift in average oxygen isotope composition of ocean water owing to the accumulation of glacial ice on Antarctica. The shaded area shows the limit of uncertainty.

A third source of information on climates is provided by fossil soils in the Palaeocene Reading Beds. At Alum Bay, on the Isle of Wight, 'a warm climate with a marked dry season' is suggested (Buuman, 1980). Seasonality is also to be inferred from fluvial sedimentary structures in the Middle Eocene beds of the Hampshire Basin (Plint, 1983), which indicate 'a fluctuating, perhaps seasonal, discharge regime'.

The climatic deterioration at the end of the Eocene, which may have been abrupt (Buchardt, 1978; Savin, 1977) or gentle (Collinson et al., 1981), according to the type of evidence used, created a different climatic regime in the Oligocene. In the Oligocene the climate of Britain was more comparable to that of a region like the south-eastern United States, though there may have been very dry intervals, as indicated by the presence of gypsum evaporite crusts and ancient continental dunes in the Ludien and Stampian sediments from the Paris Basin (Dewolf and Mainguet, 1976). The termite fauna of the early Oligocene Bembridge Marls from the Isle of Wight confirm that, although it was cooler than the Eocene, it was none the less warmer than today (Jarzembowski, 1980).

The causes of a warm Palaeogene in Britain are both local and global. At a local scale, Britain was at a lower latitude than today, being 10–12° further south in the Palaeocene. Movements of the continents may also have affected global climates by creating a unique configuration of the continental masses. Parrish (1987, pp. 52–3) has suggested that:

> By the beginning of the late Cretaceous, global palaeogeography had entered a mode of maximum continental dispersion, and the monsoonal circulation that had dominated palaeoclimatic history during the previous 250 million years had disappeared. From this time until the collision of India and Asia re-established strong monsoonal circulation during the Miocene, continental palaeoclimates were characterized by the zonal circulation that is the fundamental component of atmospheric circulation on Earth.

Other possible contributing factors to global Palaeogene warmth include: elevated atmospheric carbon dioxide levels associated with a phase of rapid sea floor spreading and plate subduction (Barron, 1985; Creber and Chaloner, 1985) or a marked reduction in the angle of tilt of the Earth's axis, possibly by as much as 15° in the Middle Eocene (Wolfe, 1978).

The hill-top peneplain of the Neogene

It has for long been believed that the higher summits of central southern and south-eastern England show a marked degree of accordance in the height range 230–260 m (Davis, 1895). Wooldridge and Linton (1939) attributed this surface to a Mid-Tertiary cycle of subaerial peneplantation which occurred after the Alpine pulse of uplift, and which planed off folded terrain.

This hill-top peneplain was thought to be largely horizontal and to transect the Alpine folds. It was also thought not to have been seriously modified by any subsequent warpings except near the coastline of East Anglia bordering the North Sea basin. Given that they believed that the Alpine folding spasm occurred in the late Oligocene and early Miocene, they attributed the peneplain to a period of earth stability in the late Miocene and early Pliocene.

One classic area which Wooldridge and Linton identified for this feature was in the Chalk uplands of Wessex (1955, p. 38):

> Standing anywhere in the south central part of Salisbury Plain one may look southward across the Wylye valley to the long even crest-line of the Great Ridge. From such a viewpoint the ridge appears to have the character of a plateau, its summit surface being separated by a sharp topographic break from the valley slopes and spurs of lower levels. It suggests strongly by its appearance that its upper surface belongs to quite another morphological cycle than that represented by the smooth, well modelled spurs and slopes that descend towards the modern valley. This upland comprises one of the largest and most instructive surviving remnants of the Mio-Pliocene peneplain.

They traced the feature and believed it was continuous across southern England from the margins of the London Basin to the western limits of the Cretaceous uplands. Wooldridge (1952) noted that a plane 800 feet (244 m) above sea-level would pass close to many of the summits of southern England; with some a little above its level (e.g. Leith Hill, Blackdown and Hindhead) and some just below it (e.g. Crowborough Beacon). He also believed that the Cotswold summits (e.g. Cleeve Hill and Broadway Hill) might also be remnants of the great mid-Tertiary plain.

Others identified the surface in other parts of Britain. Clayton (1953) correlated it with the '1000-foot' (300 m) surface of the southern Pennines, and Brunsden (1968) with the lower surface of Dartmoor (230–280 m). Correlation was based on height, and the possibility of continuing tectonic differentiation between different parts of England and Wales was largely ignored.

More recently some doubt has been thrown on this interpretation, largely because current work recognizes that the tectonic evolution of southern Britain in the Tertiary was complex (see p. 26) and that the concept of a brief and intense 'Alpine storm', responsible for the main fold-structures, is an oversimplification. Thus Small (1980) postulates that contemporaneous diastrophism and denudation may have occurred, while Jones (1980, p. 27) argues that: 'The increasingly numerous reports of the survival of Palaeogene sediments and surfaces on the higher chalklands casts doubt on both the efficacy of Neogene subaerial erosion and the concept of a Mio-Pliocene peneplain'. Amongst such sediments are the remnants of silicified sands and gravels that occur widely in south-east England as sarsen stones and Hertfordshire Puddingstone (Summerfield and Goudie, 1980; figure 1.7). Jones (1980, p. 27) concludes: 'It is now abundantly clear that stratigraphic, sedimentological and morphological arguments all support the abandonment of the Mio-Pliocene peneplain concept and its replacement by a model in which Neogene subaerial denudation led to the stripping away of the Palaeogene cover and incision into the underlying Mesozoic rocks'.

The Calabrian marine surface of Plio-Pleistocene times

The final upland surface type is that attributed to marine trimming at the very end of the Tertiary. Wooldridge and Linton (1955) identified a dissected bench, particularly in the Chilterns and the North Downs, cutting across the Chalk. Preserved on this bench they found various marine sediments at heights of around 170–200 m. These sediments were dated originally as Pliocene but subsequently were attributed to the early Pleistocene. Furthermore, Wooldridge and Linton found certain areas where the drainage appeared to be discordant to structure and postulated that this was a result of the obliteration of the pre-existing relief and drainage by the marine incursion. They mapped the extent of the transgression partly by the presence of such superimposed, discordant drainage (figure 1.8).

Controversy surrounds the extent, date and geomorphological significance of this event (Jones, 1980). Current palaeontological thinking suggests that the marine fossils on the bench (e.g. from the Lenham Beds) may be of greater age than envisaged by Wooldridge and Linton. Moreover, Jones believes that much of the discordant drainage is the result of antecedence rather than superimposition, and that rivers were able to adapt to long-continued pulsed tectonism (Jones, 1974). In south-east England he believes that the so-called Calabrian transgression 'did nothing more than cause erosional embellishments' (p. 154). Likewise, Green (1974) found no evidence that in Wessex the transgression significantly modified earlier Tertiary landforms.

Furthermore, in a series of papers, Moffat and co-workers (Moffat et al., 1986; Moffat and Catt, 1986a, b) dissected Wooldridge and Linton's evidence for the Calabrian surface in the

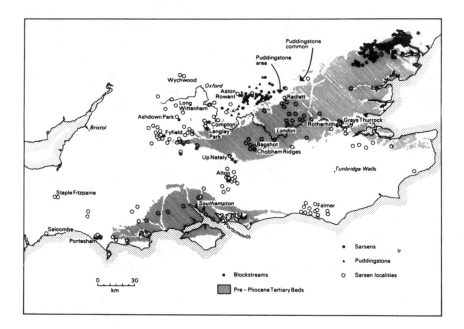

Figure 1.7 Sarsen stones, thought to be relicts of silcrete duricrusts, occur widely in southern England both in and on the margins of basins of Tertiary deposition (after Summerfield and Goudie, 1980, figure 1).

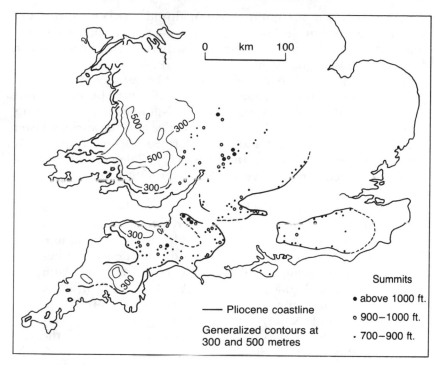

Figure 1.8 The distribution of Pliocene coastlines and ancient summit levels in Wales and southern England according to Wooldridge (1952, p. 298).

Chilterns – one of the areas where it was held to be most prominently developed. Moffat et al. (1986) showed that, although a bench existed, it was partly structural and partly due to a phase of marine erosion prior to the deposition of the Reading Beds and they concluded that 'It has no value as erosional evidence for a Plio-Pleistocene marine transgression' (p. 105). Moffat and Catt (1986a) examined the dry valley net in the Chilterns and found no evidence that drainage had been superimposed though a cover of Plio-Pleistocene marine sediment, finding that 'Drainage accords equally well with structure on "the marine platform" and "Mio-Pliocene peneplain"' (p. 178). Finally, Moffat and Catt (1986b) undertook a petrological re-examination of some of Wooldridge and Linton's marine deposits and found that, although some were indeed marine, others were fluvial in origin. Futhermore they cast doubts on the belief that the marine bench was at the same height over most of southern England, and concluded: 'The few outcrops of marine sediment are not uniform in altitude, as Wooldridge and Linton claimed, but decline towards the northeast, suggesting that considerable subsidence occurred in northern parts of the London Basin during the Quaternary' (pp. 244–5).

However, in other parts of Britain there are clearly defined platforms which truncate geological strata that may be attributable to the early Pleistocene transgression. Notable here is Brown's (1960b) '600 ft platform' (c.180 m) in Wales and the various lower platforms (e.g. at 120, 90 and 60 m) formed during progressively lower transgressions later in the Pleistocene. The current thinking on this has been summarized by Embleton (1984, p. 123):

> The highest level of about 200 m may be tentatively correlated with the Calabrian (Red Crag) marine stand recorded in south-east England. This correlation, however, based simply on altitudinal equivalence, implies widespread tectonic stability since that date, a claim which is now being questioned. Another problem is the great width of the platform in some areas (40 km in Anglesey and Arfon) . . . Remarkably effective wave-cutting is implied, especially if the resistance of the Precambrian and other strata is taken into account. Many workers have thus been forced to consider the possibility that the wide platforms of Anglesey, Pembroke-shire, Gower and Glamorgan were partially prepared for Cenozoic wave action by much older geological events . . . It is therefore possible that parts at least of the platforms may be early Triassic pediments.

Tertiary plate tectonics

Since the 1960s, the adoption of plate tectonics as a basic paradigm in earth science in general, has led to a reappraisal of the importance of continental drift and orogeny (mountain building) for geomorphological evolution. To understand the Tertiary evolution of the British landscape it is necessary to appreciate the position of the British Isles with respect to two main influences: the opening of the Atlantic and the uplift of the European Alps (Naylor and Shannon, 1982).

Prior to 200 million years ago (late Triassic) the North Atlantic did not exist as an ocean, and the continental plates of Western Europe, Greenland and North America formed one single plate (figure 1.9). During the Middle Jurassic an embryonic plate margin began to develop between Europe and North America, generating new oceanic crust over the entire length of the southern North Atlantic, and causing rifting to occur notably in the North Sea Basin (Hay, 1978). By the start of the Palaeocene sea-floor spreading had advanced sufficiently that an opening was appearing between Ireland and Greenland.

The scale of the Mesozoic rifting, which produced faulted troughs, horsts and platforms, infilled with wedges of sediments of Permian to Cretaceous age up to 10 km thick, was impressive. The North Sea Basin forms part of a large complex system, extending from Hatton Bank in the north west to Central Poland in the south east and from the Pre-Alps in the south to the Lofoten Islands in the north. To the west of the British Isles, the Hatton Bank, the Rockall–Hatton Trough, the Rockall Bank, the Rockall Trough, the West Irish Platform, the Porcupine Bank, Porcupine Trough and the South West Irish Platform form another important system of platforms and sediment-filled troughs related to sea-floor spreading and continental margin evolution (Whiteman et al., 1975).

This phase of plate separation was accompanied by a period of marked crustal extension over much of Britain and Ireland which caused a severe spasm of intrusion and extrusion of igneous material, which lasted from 61 to 52 million years ago (Dewey and Windley, 1988) (figure 1.10). Flood basalts were erupted from Hawaiian-type volcanoes to give lava sheets, notably on Mull (where they are some 2000 m in thickness) and in Antrim. Igneous intrusions developed, including the Lundy granites of the Bristol Channel, which have been dated to 52 million years ago. North-west to south-east trending dyke swarms were also intruded, and extended as far south as Lundy, Snowdonia and the north-west Midlands. Most of this activity had ceased by the early Eocene, for as the Atlantic opened still further igneous activity shifted westwards and became centred upon the location that was eventually to become Iceland. Rifting largely ceased beneath the North Sea and was replaced by gentle subsidence. This subsidence, which enabled the accumulation of much sediment (figure 1.11) and which was in part caused by sedimentary loading, can be explained in terms of the cooling (thermal relaxation) of the upper mantle that underlay the Jurassic and Cretaceous rift system

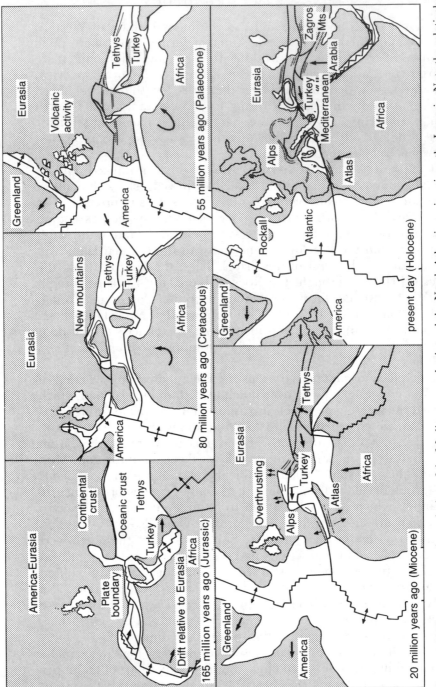

Figure 1.9 The plate tectonic evolution of the Mediterranean basin and the North Atlantic region since the Jurassic. Note the relatively late opening of the North Atlantic and the associated volcanic activity in the Palaeocene (after Dunning et al, 1978, figure 93).

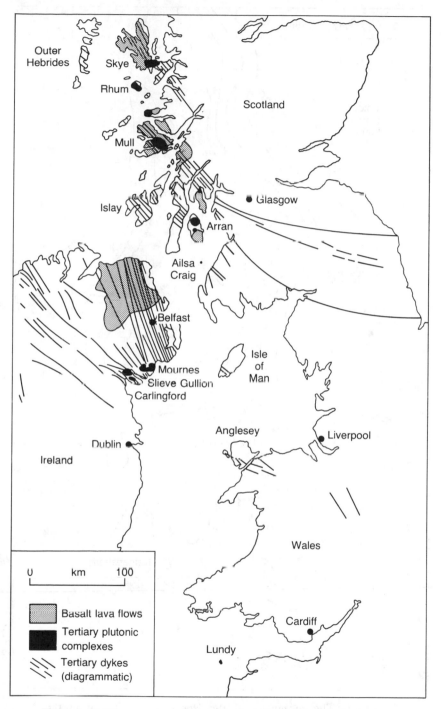

Figure 1.10 The opening of the North Atlantic produced a major spasm of igneous activity in the British Isles leading to the intrusion of plutons (including the Isle of Lundy), the injection of swarms of dykes and the extrusion of expansive basalt lava flows.

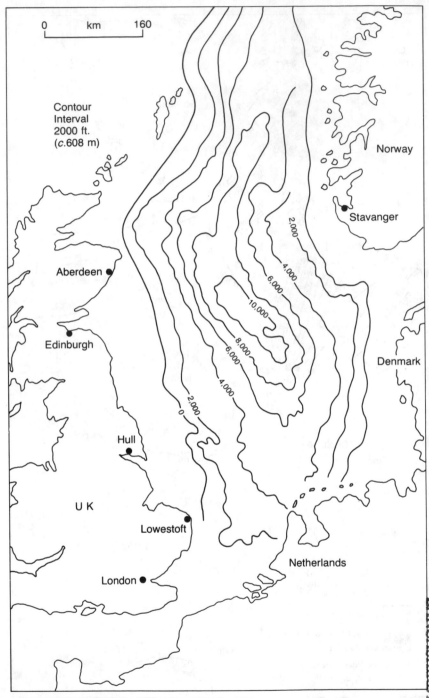

Figure 1.11 The subsidence of the North Sea basin during the Tertiary and Quaternary has enabled the accumulation of large thicknesses of sediment on top of the Chalk. In the case of the central part of the basin there are over 10,000 feet (*c*.3,000 m) of material (modified from Lovell, 1986, figure 8.3).

(Sclater and Christie, 1980). The cooling caused a corresponding density increase of the crustal rocks, resulting in further subsidence (Ziegler, 1978).

During the late Mesozoic, the collision of Europe and Africa created compressional forces that caused a major orogenic phase. This tendency reached its climax during the late Eocene and early Oligocene (Owen, 1976). The deformations of strata which took place in Britain have often been seen as the ripples that occurred on the margins of 'the great Alpine storm'. In addition, widespread faulting occurred, frequently rejuvenating pre-existing ancient structures and causing broad warpings of the crust. However, it is not entirely clear that these foldings, faultings and warpings necessarily have an 'Alpine' pedigree. As Anderton et al. (1979 p. 255) suggest: 'They could reflect crustal readjustments following the Palaeocene departure of Greenland due to sea floor spreading. In any case, "Alpine" effects in Britain were probably magnified as a result of early Tertiary Atlantic opening.'

It is also evident that the tectonic activity was more continuous than often believed and that pulsed tectonism occurred from the late Cretaceous throughout the Palaeogene to the mid-Tertiary (George, 1974; Jones, 1985). This view is in marked contrast to Wooldridge and Linton's (1955) interpretation of structural evolution in southern England, which envisaged the Tertiary as characterized by tectonic stability, with the exception of two periods of disturbance: Upper Cretaceous emergence and a brief late Oligocene–early Miocene 'alpine' spasm.

Some seismic activity continues in Britain at the present day and, in 1580, the London earthquake reached a magnitude of 6.2–6.9 on the Richter scale (Neilson et al., 1984), although 'No recorded British earthquake has ever been accompanied by perceptible crust-displacements at the surface' (Davison, 1924, p. 7). Davison lists 1,191 earthquakes in Britain between 974 AD and 1924 AD; although the majority of these occurred in Scotland. He identified certain areas (figure 1.12) that were relatively free of them: north-east England, central Wales (including Shropshire and Cheshire) and central southern England between Oxford and the south coast. He also noted a clear association with the existence of major faults. This has been confirmed recently in Wales (Blenkinsop et al. 1986), where a relationship has been proved between earthquake epicentres and such major fault structures as the Bala Fault and the Swansea Valley disturbance.

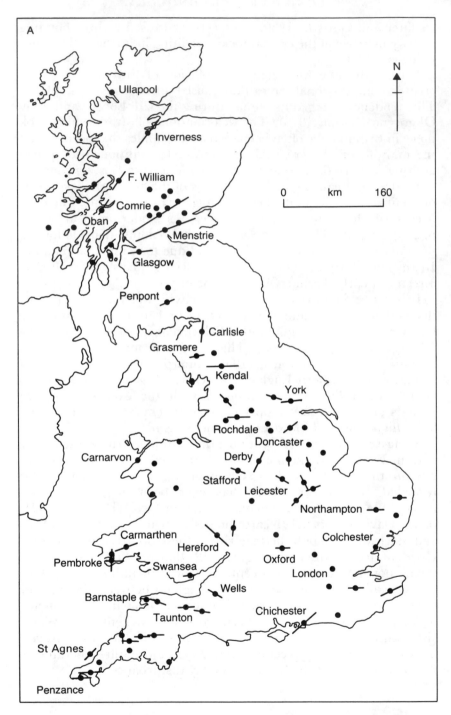

Figure 1.12 (A) Davison's early map of the distribution of British earthquakes (from Davison, 1924, figure 94); (B) the epicentres of

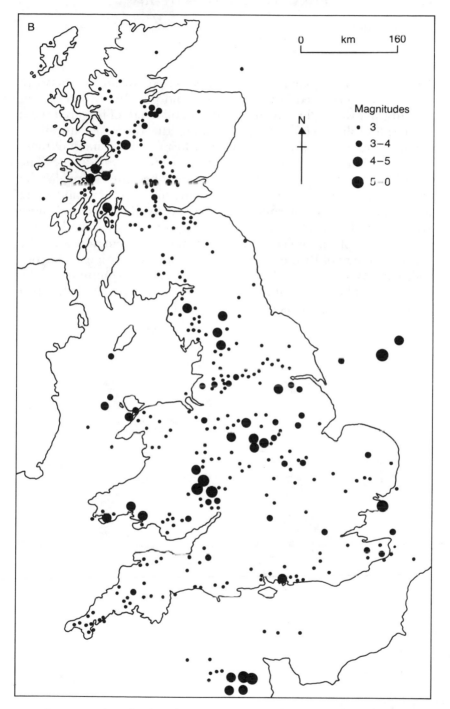

known earthquakes based on more recent analyses by the Institute of
Geological Sciences (from Perry, 1981, figure 8.1).

Conclusion

That landscape planation and associated drainage network development occurred in the Tertiary (and possibly earlier) is not in doubt. However, in spite of a century and a half of research there is considerable uncertainty surrounding the three key questions posed by Belbin (1985): how were they formed? how many of them are there and where are they? when were they formed? The situation may develop further, and possibly become clearer, as the implications of plate tectonic movements are explored, and more sedimentary and stratigraphic data become available from offshore areas. Denudation chronology as an approach may have become less fashionable in recent years, but the fact remains that the geomorphology of Britain demonstrates a long evolutionary history and, with Wooldridge (1958, p. 34), one cannot countenance the idea that: 'These beautifully integrated landscapes had dropped ready-made like gobbets of manna from heaven.'

2

GLACIATION

Climatic decline: glacials and interglacials

As we saw in chapter 1, the climate of the world in general, and of Britain in particular, was very different during the Tertiary (which started with the Palaeocene some 65 million years ago) to what it is today. The North Atlantic region in the early Tertiary was characterized by a widespread tropical or subtropical rainforest vegetation and in the London and Hampshire basins of southern England *Nypa* palm (currently typical of Malaysia) and *Avicennia* (a species of mangrove) swamps were present. Deposits from the Bovey Tracey fault basin in western England, which are probably of pre-Miocene age, contain many tropical-type plants, including *Osmunda, Calamus, Ficus, Symplocos* and *Lauras*. Soils and weathering horizons in the Tertiary basalts of County Antrim, in Ulster, are reminiscent of the classic laterite deposits of South India. Tertiary lateritic weathering has also been identified in Devon (Isaac, 1983).

However, as the Tertiary progressed, the so-called 'Cainozoic climate decline' set in. The causes of this trend are still not fully understood, but it appears to be associated with the break-up of the ancient super-continent of Pangaea into the individual continents we know today. About 50 million years ago, as a result of plate movements, Antarctica separated from Australia and gradually shifted southwards into its present position centred over the South Pole. At the same time, following the opening of the Atlantic Ocean, the continents of Eurasia and North America moved towards the North Pole. As more and more land became concentrated in high latitudes so increasingly large ice caps could develop, causing surface reflectivity to increase and thereby probably causing climate to cool all over the world.

The study of sediment cores from the floor of the North Atlantic Ocean on the Rockall Plateau (Shackleton et al., 1984) indicates that substantial accumulation of iceberg rafted debris occurred around 2.4 million years ago (in the late Pliocene) and this, they believe, must correlate with the first fully glacial environment in Britain. There is considerable supporting evidence for a sharp climatic deterioration at about this time from many other parts of the world, promoting extensive mid-latitude glaciation (Goudie, 1983, pp. 26 and 30).

The following 2.4 million years (which include the 1.6 million years of the Quaternary) were a time of remarkably extreme, frequent and rapid climatic changes, as, under the stimulus of perturbations in the Earth's orbit (the Milankovitch cycles) a whole suite of cold episodes (glacials) and warmer intervals (interglacials) occurred. In all there have probably been about 17 glacial/interglacial cycles in the Pleistocene itself, with nine of them occurring in the last 0.9 million years. The warm interglacials (broadly comparable in climate to the present day situation) were relatively short-lived, only constituting about 10 per cent of the time. They seem to have had a duration of the order of 10^4 years, while a full glacial cycle seems to have lasted around 10^5 years. Thus conditions such as those we experience today have been atypical of the Quaternary as a whole. Many of our present landforms have been shaped by the *cold* phases of the last 2.4 million years.

Much of the evidence for the frequency of Pleistocene climatic changes has come from the relatively intact and continuous record of sediment accumulation on the sea floors. The evidence on land is much less clear, especially because the remains of previous events have been obliterated by subsequent erosion. The further back in time one goes the more imperfect the record becomes. Furthermore, reliable absolute dating of the terrestrial record is extremely difficult. Thus interpretation of British glacial and interglacial sequences is fraught with controversy so that in spite of over 150 years of investigation there is little unity of view or certainty of evidence (Bowen, 1978).

The classic view (see, for example, Mitchell et al., 1973) is that there have been three major glacials in Britain during which ice caps covered large areas of the country. The earliest of these (called the Anglian) is presumed to correlate with the Elster Glaciation of

continental Europe. Its extent is not delimited by any well-marked
end-moraine system but by dispersed till deposits, which in the
case of the Plateau Drift deposits of the Oxford Region are often
little more than scattered, thin, accumulations of erratic pebbles,
the origin of which is far from proven. A good sequence of Anglian
materials occurs at the type site at Corton Cliff, near Lowestoft,
Suffolk. The Anglian glaciation may have been the most extensive
to affect the British Isles (figure 2.1), reaching as far south as
Oxford and south Essex.

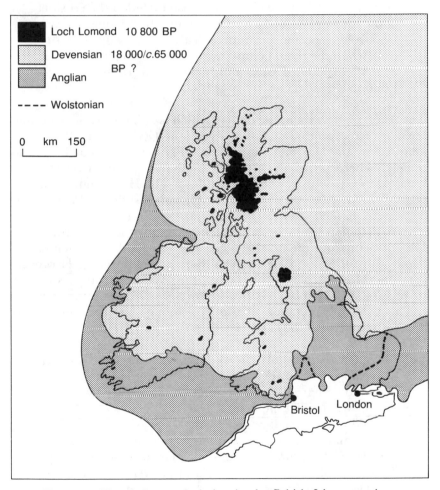

Figure 2.1 The limits to glaciation in the British Isles at various
stages in the Pleistocene. Note that the existence of the Wolstonian
glaciation is the subject of controversy.

The Anglian was undoubtedly a very important event in the geomorphological history of Britain for it produced some of the greatest changes in the evolution of the landscape. As Bowen et al. (1986, p. 308) put it:

> In particular it is the time when the drainage of East Anglia changed from an extensive series of northeastward sloping river terraces, to a glacial landscape composed primarily of till with a radial drainage pattern. The period of the Anglian Glaciation also included the erosion of the lowland of the Severn basin, and hill ranges like the Cotswolds became prominent features. Finally it is the time when the river Thames was directed from the Vale of St. Albans and Essex into its present route through London.

The Anglian glaciation, the date of which is still uncertain, but may have occurred at around 450,000 years BP (Eyles et al., 1989) was followed by the Hoxnian Interglacial. The type site is at Hoxne in Suffolk, where in a hollow in the Anglian till are lacustrine and fluvial deposits with an interglacial flora. The Hoxnian in turn was possibly followed by the second main glacial episode, the Wolstonian cold stage, named after the type site at Wolston, in Warwickshire. The Wolstonian ice sheet, if it existed, extended as far south as the Isles of Scilly and the northern coast of the south-west peninsula, depositing till at Fremington near Barnstaple (Devon). However, considerable controversy surrounds the Wolstonian type site and its precise stratigraphic position (Sumbler, 1983; Shotton, 1983).

The new interpretation of the Wolstonian has been expressed forcefully by Rose (1987, p. 1):

> a regional study of the Baginton/Lillington Gravels, which form the lower part of the Wolstonian at its type site, show that these sediments can be traced to East Anglia where they lie *beneath* glacial deposits of Anglian age. Thus, the sediments at the type site, and other 'Wolstonian' deposits in Midland England were found *prior to*, and *not after* the Anglian glacial. Consequently, the extent of glaciation in England between the Anglian and Devensian stages is in need of revision, and the use of the term 'Wolstonian' to describe the glaciation at this time is no longer appropriate.

There is also considerable controversy surrounding the status of the type site of the next interglacial, the Ipswichian, named after a site at Bobbitshole near Ipswich (Bowen, 1978, p. 36). However, we reach firmer ground when we come to the last major glaciation in Britain, the Devensian. At its maximum extent it covered over two-thirds of the present land area of the country, smothering almost all of Scotland, a very substantial area of Ireland and Wales, and large parts of northern England. Ice extended down the coast of Holderness, Lincolnshire and into the Hunstanton area of north Norfolk. Its maximum extent was reached some 18–17,000 years ago. However, there was a glacial re-advance, following its wastage, which lasted for about 450 years from 11,000 years ago. This advance, the Loch Lomond Stadial 'was the last time that England experienced the presence of glacier ice, and the last time that it experienced rapid landscape change before coming covered with the protective woodland of the Flandrian Interglacial' (Bowen et al., 1986, p. 311).

The classic view just outlined is the subject of continuous revision and discussion. There may well have been earlier glacial events in the Pleistocene, the evidence for which is fragmentary. The present position has been summarized by Bowen et al. (1986, p. 299):

> Extensive lithological evidence for glaciation, with a sound stratigraphic basis, has been recognized in two stages of the English Pleistocene: the Anglian and the Devensian. Additionally, less extensive or indirect lithological evidence, often with a less secure stratigraphic basis, indicates that glaciation occurred in at least a further six episodes.

There is evidence preserved on the floor of the North Sea for an early glacial episode, dated prior to 1.8–2.1 million years BP (Sutherland, 1984).

What was the maximum extent of glaciation?

Whatever the precise status of the Anglian and Wolstonian glaciations proves to be, a highly important consideration for understanding of landforms is the maximum extent that was attained by glaciation of whatever date. This became an issue of some controversy in the 1970s, largely as a result of some arguments put forward by Kellaway (1971) and Kellaway et al. (1975).

Having identified that ice had crossed the Severn Estuary from Wales and entered Somerset, the evidence for which is provided by erratic material exposed in new motorway sections (Hawkins and Kellaway, 1971), Kellaway (1971) went on to speculate that ice had reached Salisbury Plain (providing the stones for Stonehenge in the process) and overspilled into the Hampshire Basin; though, as Green (1973) pointed out, there are no glacially derived erratics in the Hampshire Basin rivers to substantiate this view. Kellaway et al. (1975) later went further still, proposing that an Anglian ice sheet moved up the English Channel from the west and affected parts of the South Downs. If such a model was correct then the bulk of the landscape of southern Britain was glaciated.

The 'evidence' put forward by Kellaway for the occurrence of very extensive glaciation includes the presence of such features as: the enclosed 'deeps' or 'fosses' of the English Channel floor (which he regarded as being due to subglacial scour); valley bulges and landslips (which he regarded as being the result of glacial loading and oversteepening); Clay-with-flints (regarded as decalcified till); erratic materials in Chesil Beach, Sussex raised beaches and at miscellaneous sites (e.g. Stonehenge); and the U-shaped dry valleys of the Chalk downlands. However, as Jones (1981, p. 50) points out: 'There is much conjecture and many circular arguments in this hypothesis, especially as most of the landforms and deposits mentioned have been explained adequately by geomorphological processes other than glacial ice.'

Another attempt to prove that the normally accepted glacial limits were too northerly was made by d'Olier (1975) as a result of investigations of enclosed depressions off the Essex coast. He suggested that the depressions were ice scour phenomena and that an ice sheet extended southwards across the present Thames Estuary to Thanet, possibly penetrating the Straits of Dover. However, a tidal scour origin for such depressions is a more widely held explanation.

Uncertainty remains about the precise limits of glaciation in the London Basin, where there are some erratic-rich pebbly clays which have been tentatively interpreted as the deeply weathered remnants of tills laid down by an early glacial advance. This material, called 'The Chiltern Drift', occurs on the Chiltern

backslopes of Hertfordshire, the South Hertfordshire Plateau and
the Epping Forest Ridge. Baker and Jones (1980) suggest that it
may be a composite lithostratigraphic unit, part solifluction deposit
and part deeply weathered early till.

Bearing all these uncertainties in mind, the maximum extent of
glaciation ran through northern parts of the Isles of Scilly (Mitchell
and Orme, 1967; Coque-Delhuille and Veyret, 1989), the north
coast of the south-west Peninsula, the area just to the south of
Bristol, the Oxford Region and Essex.

The Devensian cold stage

The Devensian is the best preserved and best dated of the various
cold stages that affected Britain in the Pleistocene. It provides scope
for detailed environmental reconstruction within the framework of
some tolerably reliable dates and can act as a model for what
conditions may have been like in early glacial phases. However,
even the Devensian record is based on a number of key sites, each
of which contains a palaeoenvironmental record for only a portion
of its total duration. At no site is there a continuous sequence of
deposits preserved that spans the whole of the Devensian cold stage
(Lowe and Walker, 1984, pp. 302 ff).

The base of the Devensian in Britain is not easy to define.
Cooling from the last Interglacial (Ipswichian?) probably occurred
at around 110,000 years ago and there was probably a series of
stadials and interstadials in the early Devensian. Whether glaciers
existed in Britain at this time is a matter for conjecture; they may
have occurred in upland areas. However, the presence of traces left
by ground ice mounds and involutions suggests permafrost may
have been developed; while pollen analysis from a number of sites
indicates an almost complete absence of trees from the landscape of
south-eastern England. Around or perhaps even before 60,000 BP
an interstadial occurred – the Chelford Interstadial – named after
the type site near Congleton in Cheshire. A birch, pine and spruce
woodland developed over large areas of central and southern
Britain, similar to the modern boreal forest of southern Finland.
The climatic regime was continental with July temperatures only
slightly below those of today, but with extremely low January
temperatures (possibly less than $-10°C$). Following the climatic
amelioration of the Chelford Interstadial, a relatively long period of

predominantly cold conditions prevailed. Lowe and Walker remark (1984, p. 316) of the English Midlands at that time:

> The coleopteran evidence, in association with the limited palaeobotanical data, indicates a climatic regime of arctic severity, similar to that of the north-east coast of European Russia today. This would imply a mean July temperature of around, or perhaps even below 10°C, winter temperatures ranging between −4°C and −8°C, an annual precipitation of 400–600 mm with more than 50 cm of winter snowfall . . .

There then followed a relatively warm interval, the Upton Warren Interstadial, with a thermal maximum between approximately 42,000 and 43,000 years ago. The landscape of lowland Britain was one of open grassland with a climatic regime similar to that of north-central Europe today. July temperatures were in the region of 17° to 18°C, January temperatures were around 0°C to −2°C and a precipitation of 550 mm per year is implied.

By shortly after 40,000 years BP severe climatic conditions returned, producing a barren tundra landscape from which trees were absent. It was not until the late Devensian, probably between 30,000 and 25,000 BP, however, that substantial glacier build-up occurred, leading to the formation of a very extensive ice cap. By 22,000 BP summer temperatures were below 10°C and the coldest winter month around −16°C or below (Atkinson et al., 1987). Boulton et al. (1977) used miscellaneous geomorphological information to suggest that, at its maximum, the ice sheet consisted of a series of domes centred on the main dispersal centres, the largest of which (over central Scotland) had a summit height of over 1,800 m (more than 450 m above the summit of Ben Nevis). This would imply local ice thicknesses in the last ice sheet of well in excess of 1,200 m. The precise lateral limits of the ice sheet are not entirely clear, especially in the west and the east. It is probable, however, that it extended only a short distance offshore along the east and west coasts (Sutherland, 1984), and the British and Scandinavian ice sheets may not have coalesced at this time (Sejrup et al., 1987). Cameron et al. (1987) report that the Devensian till on the North Sea floor has a well-defined eastern boundary, mostly less than 100 km offshore and that, for its part, Scandinavian till does not extend beyond the western margin of the Norwegian Trench.

The glaciers reached their maximum extent at $c.18000$ years ago and, with possible short interruptions, retreated rapidly thereafter, so that the ice sheet had largely disappeared by 13,000 BP, leaving most of the Scottish Highlands ice-free. An ice volume that may have approached 346,000 km^2 wasted away almost completely in less than 5,000 years. Rose (1985) believes that the whole of the late Devensian glacial episode (130,000–26,000 years BP) should be called 'The Dimlington Stadial' after the type site in East Yorkshire.

The Loch Lomond Stadial

The breaking up of large ice shelves from the Arctic may have caused a rapid cooling of the Atlantic Ocean, setting in train a climate deterioration (Ruddiman and McIntyre, 1981).

Cold conditions returned to Britain for a brief period at the end of the late Devensian, producing the Loch Lomond Stadial. Glaciers began to reform in many upland areas. The greatest area of ice accumulation was in the Scottish Highlands; over Rannoch Moor ice thickness may locally have exceeded 400 m. Smaller ice caps existed in the Grampians and on Mull and Skye in the Inner Hebrides. Cirque and valley glaciers were still more widespread, not only in the Scottish Highlands but in the southern Uplands of Scotland, the English Lake District, and miscellaneous mountainous parts of north Wales (e.g. Snowdonia, Cader Idris) and the Brecon Beacons. Their end-moraines are still very clearly evident in the present landscape, while beyond the ice margins severe frost action produced periglacial features (see chapter 3) which are also well preserved – summit blockfields, solifluction lobes and spreads, screes, rock glaciers, protalus ramparts, etc. Although short lived, this Loch Lomond Stadial was an event of immense geomorphological importance (Sissons, 1980).

Gray (1982) has made a detailed study of the effects of the Loch Lomond advance in Snowdonia. He estimates that there were 35 corrie glaciers at that time. In total they only covered 17.5 km^2 so their average size was small ($c.0.5$ km^2). The biggest glacier was the Llydaw (area $= 4$ km^2, thickness over 250 m). The equilibrium firn line (mean 600 m above sea level) rose from 450 m in the south west of Snowdonia to 700 m in the north east. Studies of glacial erosional forms suggest that the Loch Lomond Stadial glacier had less erosional power in Snowdonia than those of the Devensian glacial maximum (Sharp et al., 1989).

Comparable work in the Lake District (Sissons, 1980) revealed the presence of 64 glaciers at the time of the Loch Lomond advance. In total they covered 54.55 km^2 (three times the area of those in Snowdonia) and had a mean size of 0.85 km^2. The mean altitude of the firn line, at 540 m, was, as might be expected, slightly lower than in Snowdonia.

The north side of the Brecon Beacons escarpment in South Wales also has some distinctive, fresh end moraines which Walker (1980, 1982) has attributed to Loch Lomond Advance times.

The distribution of glacial erosion

An attempt to map the principal large-scale pattern of erosion in Britain has been made by Boulton et al. (1977). (figure 2.2). They employ five empirically derived categories of modification by ice (table 2.1) from Zone 0, which is virtually unmodified, to Zone IV, where the entire landscape has been shaped by ice.

As one might anticipate, the zones of highest erosional intensity are the Lake District and the mountains of north-west Wales; both areas where there are major cirques, glacial troughs, scoured surfaces, and examples of glacial diffluences and severe drainage modification.

The Lake District, because of its high elevation and westerly situation, would have received a high level of precipitation, giving a thick and active ice sheet, with numerous outlets and steep gradients. Linton (1957) suggested that ice largely remade the drainage alignments of the area, with a radial pattern of glacial troughs developing as a result of radial flow in all directions from the main ice-cap centre. There are also all the classic features of corrie basins with tarns, of arête ridges (e.g. Red Tarn and Striding Edge on Helvellyn), and of overdeepened valley troughs (e.g. the northern basin of Windermere descends to − 21 m OD). The degree of glacial modification is not, however, as severe as that experienced in north-west Scotland. Some preglacial erosion surfaces still survive (King, 1976, p. 102) and glacial erosion has been essentially concentrated in the valleys, where ice flowed faster

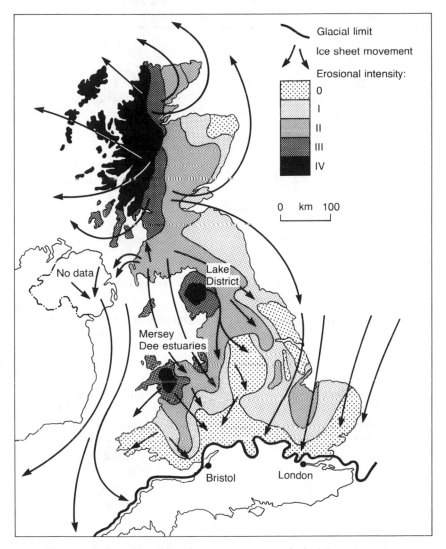

Figure 2.2 Patterns of ice-sheet movement and glacial erosion. For explanation of the five zones of erosional intensity see table 2.1 (modified from Jones, 1985, figure 1.23).

and thicker. Much of the higher, flatter parts of the interfluves have probably been much less altered by glacial erosion.

The mountains of North Wales form the other classic area of severe glacial erosion, and their glacial features were an important source of evidence for those who championed the existence of the

Table 2.1 Zones of glacial erosion

Zone	Lowlands	Uplands
0	No erosion Head on weathered rocks and slopes Outwash in concavities Rare occurrences of till on weathered rock	No erosion Outwash on valley floors Solifluction deposits on slopes, boulder fields and tors on divides
I	Ice erosion confined to detailed or subordinate modifications Concavities drift mantled but convexities may show some ice moulding Occasional roches moutonnées Ice-scoured bluffs in favourable locations	Ice erosion confined to detailed or subordinate modifications Suitable valley slopes ice steepened Entrenched meanders and spurs converted to rock knobs Interfluves still commonly Zone 0
II	Extensive excavation along main flow-lines so that concavities may be drift free or floored by outwash or post-glacial deposits. Isolated obstacles may be given ovoid or cutwater forms if of soft rock, or crag-and-tail with associated scour troughs if hard	Conversion of preglacial valleys to troughs common, but usually confined to those of direction concordant with ice flow Some diffluence; transfluence rare Interfluves may be Zone I or even Zone 0, and separated from troughs by well-marked shoulders
III	Preglacial forms no longer recognizable but replaced by tapered or bridge interfluves with planar slopes on soft rocks, and by rock drumlins and knock-and-lochan topography on hard rocks	Transformation of valleys to troughs comprehensive giving compartmental relief with isolated plateau or mountain blocks. Zone 0 may still persist on interfluves at sufficiently great heights.
IV	Complete domination of streamlined flow forms even over structural influences	Ice moulding extends to high summits Upland surfaces given knock-and-lochan topography (sometimes of great amplitude at lower levels) Lower divides extensively pared or streamlined

Ice Age a century and a half ago. The pronouncements of Charles Darwin on valley glaciation (1842, p. 188) are typical:

> I cannot imagine a more instructive and interesting lesson for any one who wishes (as I did) to learn the effects produced by the passage of glaciers, than to ascend a mountain like one of those south of the upper lake of Llanberis, constituted of the same kind of rock and similarly stratified, from top to bottom. The lower portions consist entirely of convex domes or bosses of naked rock, generally smoothed, but with their steep faces often scored in nearly horizontal lines, and with their summits occasionally crowned by perched boulders of foreign rock. The upper portions, on the other hand, are less naked, and the jagged ends of the slaty rocks project through the turf in irregular hummocks; no smooth bosses, no scored surfaces, no boulders are to be seen, and this change is effected by an ascent of only a few yards.

The great American geomorphologist W. M. Davis was equally struck by the transformation of Snowdonia that had been achieved in the Glacial Period (Davis, 1909, p. 281): 'An excursion around Snowdon . . . led me to the conclusion that a large-featured, round-shouldered, full-bodied mountain of pre-Glacial time had been converted by erosion during the Glacial Period – and chiefly by glacial erosion – into the sharp-featured, hollow-chested, narrow-spurred mountain of today.' He believed that the scalloped cwms, rock steps, deep basins, hanging valleys and truncated spurs, were proof of the role of glacial scour.

However, in the cases of both North Wales and the Lake District there seems to be a broad-scale lithological control of the best developed glacial erosional forms, for ancient volcanic rocks appear to give rise to the most spectacular features (Sparks, 1971, p. 112). In the case of Snowdonia, Ordovician rhyolites play this role; in Cader Idris it is Ordovician acid volcanics; in the Lake District it is the andesitic lavas and tuffs of the Ordovician Borrowdale Volcanic Series. Sparks (1971, p. 278) stresses this point further in trenchant style by discussing the relationship between different Ordovician lithologies and glacial landform development in Wales as a whole:

> The greatest density of igneous rocks is reached in the Snowdon area and in the broken escarpment surrounding

Plate 2. The Snowdon massif is one of the classic areas for finding evidence of glacial erosion in Wales. It is noted for its deep cwms, lakes and arêtes. (Cambridge University Collection: copyright reserved)

the Harlech Dome. A much lower density is present in the Berwyn Dome and in the inliers around Plynlimmon, the latter being almost entirely greywacke in type. Although glacial landforms occur in the Berwyns and again at Plynlimmon, just as they do in Snowdonia, they are not really comparable. There is a corrie on the northern side of Plynlimmon – it even has rocky cliffs fringing it on the mountain side, but it would be fair to describe it as pathetic compared with the developments on the southern side of the Nant Ffrancon valley, with the corries that bite into Snowdon, or with Llyn Cau between the northern and southern ridges of the Cader Idris block. Not only that: Plynlimmon is generally soft and rounded in relief with well-developed peat bogs slightly below the maximum elevations. Cader Idris, Snowdon, Tryfan and the Glyders are mountains of jagged rocks – in fact, mountains and not hills. Whether these differences are due primarily to glaciation or primarily to rocks is difficult to argue, for both process and rocks have contributed. It might be argued that glacial forms develop best on igneous rocks. It could also be argued that glacial forms are best preserved on igneous rocks. On the other hand, it could be alleged that Snowdonia was the centre of an ice sheet and that the Berwyn Dome was not. But even this explanation throws the ultimate responsibility back on to the rock type: if Snowdonia was the centre of an ice sheet, it was because it was higher; if it was higher, it was because the rocks were more resistant.

However, the scale of glacial erosion in upland Wales should not be exaggerated. Indeed, Battiau-Queney (1981) has suggested that the effects of glaciation were highly selective according to the resistance of the rocks involved; that large areas of preglacial relief and weathering profiles have survived glaciation; and that large areas of mid-Wales offer virtually no glacial erosion features at all.

Related to this idea is the extent to which glaciated valleys were or were not cut by valley glaciers for, although the traditional view is that valley glaciers have excavated U-shaped troughs, an alternative model can be envisaged. In such a model an ice sheet would cover an area like the Lake District or North Wales, and within the ice sheet there would be fast-moving streams of ice which would be frozen to the bedrock. By contrast, the thicker ice in the valleys would slide rapidly over the rock surface aided by a thin film of meltwater. Erosion would therefore be concentrated in the valleys and the contrasts between valleys and interfluves would be enhanced. Glacial erosion might therefore be highly selective according to the nature of preglacial topography (Boardman, 1988). Indeed, Boardman (p. 43) believes: 'It is important to escape from excessive reliance upon alpine models of valley glaciation. The Lake District has been repeatedly affected by ice sheet glaciation and troughs such as that at Wasdale are probably the result of such conditions. There are few of the classic features of alpine glaciation – hanging valleys, terminal and lateral moraines.'

Even if ice sheets rather than valley glaciers are accorded a major role in shaping glacial troughs, there still remains controversy over which ice sheets carried out the erosion. This is particularly the case in North Wales, where there are two distinct points of view. On the one hand some geomorphologists have proposed that the mountains maintained a subsidiary ice centre within the main Welsh ice sheet, and that the landforms of glacial erosion in Snowdonia were excavated by local ice masses radiating from that centre. The alternative view is that the main Welsh ice sheet overrode Snowdonia during glacial maxima, producing the major through-valleys of the area, while cirques were excavated by smaller local glaciers either before or after ice-sheet conditions developed. These opposing views are discussed by Gemmell et al. (1986) who, on the basis of their own analyses of the directions of glacial striae in the area favour the former view, with glacial erosion radiating out from the vicinity of the Snowdon summit.

Corrie orientations

In common with most northern hemisphere corries, those of Britain display a clearly preferred orientation. This is demonstrated in a qualitative way in figure 2.3,, which shows the distribution of corries in the Glyder range (Y Glyderau) of North Wales. There is a clear preferential orientation to the north east (Seddon, 1957).

This qualitative impression has been confirmed for North Wales as a whole by Unwin (1973), who found an overall mean orientation for 81 corrie basins of 47.5°E. (table 2.2). He believes that such a consistent overall concentration of orientations cannot be explained in simple structural terms, and that a general climatic control needs to be adopted: the redistribution of snow by prevailing south westerly winds into sheltered lee hollows; a slight tendency for snowfall to be at a maximum not on exposed south-west-facing slopes but also just over hillcrests; and, most important of all, a relative freedom from ablation on shaded north-east-facing slopes.

Confirmation of this preferred orientation is provided by

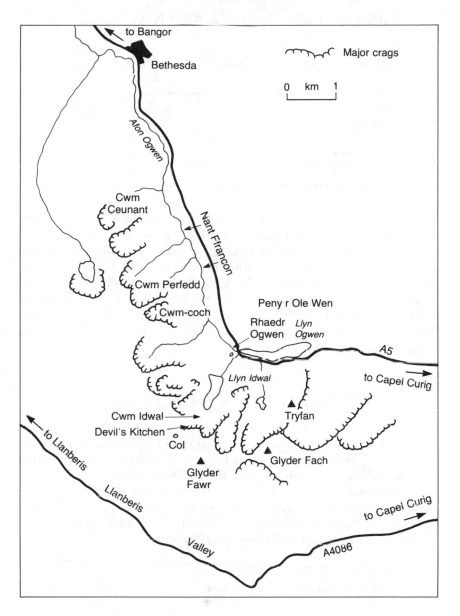

Figure 2.3 The corrie (cwm) basins of the Glyder/Nant Ffrancon area of Snowdonia. Note the preferred north-east-facing orientations.

Table 2.2 Corrie axis locations in North Wales (from
Unwin, 1973, table III)

Hill Group	No. of corrie basins	Resultant vector (°E of N)
Hebog	19	56.5
Snowdon	25	41.2
Y Glyderau	21	42.8
Y Carneddau	19	56.1
All	81	47.5

Temple's (1965) analysis of cirque distribution in the Lake District, where 53.4 per cent of features occur in the north-east quadrant, 24.7 per cent in the north-west, 19.2 per cent in the south-east, and only 2.7 per cent in the south-west. A broadly similar picture emerges for the cirques of the Old Red Sandstone escarpment (including the Brecon Beacons) in South Wales where the mean orientation is 22°E (Ellis-Gruffydd, 1977) and from Cader Idris in North Wales where features like Cwm Cau point north eastwards (Watson, 1960).

Glacial erosion in lowland areas

Although the evidence for glacial erosion in highland areas is often abundantly clear, as exemplified by such phenomena as corrie basins, roches moutonnées, striated bedrock surfaces, and over-deepened valleys, it has often been maintained that in lowland areas, by contrast, ice sheets were agents of deposition rather than of erosion. Sparks and West (1972, p. 82) have expressed a justifiable note of caution about this contention:

> It is undoubtedly true that deposition is the main effect, but the deposition of large quantities of material implies its erosion somewhere or other. The main product in lowland Britain is ground moraine and the raw material for this has not been derived from the highlands where most moraines are largely sand and gravel. In the lowlands they are predominantly clays derived from no

great distance. This can be seen in the way in which the nature of the till usually varies with the rock on which it lies, though obviously its character cannot change immediately one passes from one outcrop to another.

The composition of lowland till sheets suggests that much of it has been derived from erosion of the Lias, Oxford, Kimmeridge and Gault clays, and also of the Chalk.

Locally there is evidence of spectacular, en masse movements of material. Near Ely in Cambridgeshire, for example, there is an erratic of Chalk, Gault and Lower Greensand, measuring around 400 × 50 m and weighing 1–2 million tons. There are also huge chalk rafts in the boulder clay cliffs of North Norfolk (Banham, 1975) at such places as Trimingham, Sidestrand and East and West Runton. These large rafts, which are uniformly some 10 m thick and may have measured several hundreds of metres laterally, were probably derived from the floor of the North Sea.

However, the most convincing geomorphological indication of the power of glacial erosion in lowland Britain is possibly provided by the form of the Chalk Escarpment of the Chilterns compared to that in East Anglia (figure 2.4). To the south of the limit of glaciation the scarp has an altitude of about 250 m, whereas to the north it is broad and irregular and far lower, descending to 50 m in northern East Anglia. This suggests that the crest of the escarpment, which is also set back by around 3 km, has been lowered by the eroding ice by 100 m or more. The Fen streams, Lark, Little Ouse and Wissey, rise in an imperceptible divide some 40 km east of the Gault outcrop. 'This complete failure of a major escarpment over so great a distance . . . must be attributed to massive erosion by ice deploying from the Fen Basin' (Linton, 1963, p. 8).

The map of Boulton et al. (1977) (figure 2.2) shows one further area of severe (Zone III) lowland erosion in the vicinity of the Mersey and Dee estuaries. Gresswell (1964) postulated that these two estuaries both had forms that were atypical of the open funnel pattern of most other drowned river estuaries – the Mersey with a banana-shape and the Dee with a rectangular one. He also pointed to the fact that beneath the estuaries there were deep depressions in the bedrock surface, which lie at more than 76 m beneath parts of the Dee. He believed that the shapes and the depressions were both caused by erosive ice moving in a south-easterly direction from the Irish Sea basin into the Cheshire–Shropshire Plain.

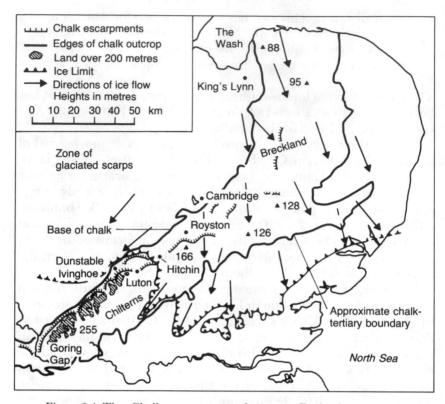

Figure 2.4 The Chalk escarpment of eastern England was very
greatly modified by Pleistocene glacial erosion. Compare the position
of the escarpment, with regard to the position of the base of the Chalk,
in the zone of glaciated scarpland with that of the scarp to the south of
the glacial line.

Glacially dammed lakes

One consequence of the presence of ice sheets and valley glaciers in
Britain was that there was wholesale blocking and modification of
drainage. One manifestation of this was the existence of sometimes
large, glacially dammed lakes.

A great proponent of the importance of such lakes was
P. F. Kendall (1902), whose paper on glacier lakes in the Cleveland
Hills had a profound effect on British glacial geomorphology.
Kendall's thesis was that ice was watertight and could thus act as a

dam, permitting the development of large lakes in valleys not actually occupied by ice. Nowhere does he refer to the possibility of drainage into or beneath the ice (Price, 1973).

Figure 2.5 shows Kendall's view of the nature of glacial lakes in the North York Moors, with a large 'Lake Pickering' and various major overflow channels (e.g. Newtondale).

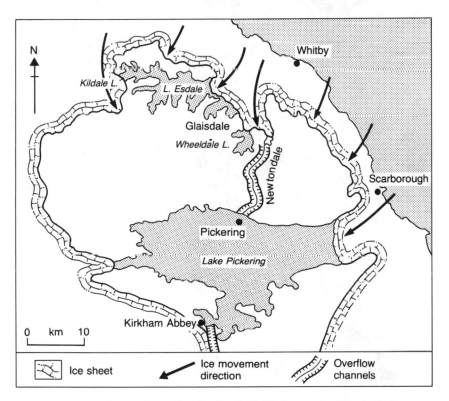

Figure 2.5 P. F. Kendall's classic glacial lakes in the North York Moors (from Kendall, 1902, p. 571). Note the Newtondale 'overflow' channel.

Subsequently, 'lakes' proliferated over many of the lowlands and vales of Britain and fertile minds began to imagine huge lakes to explain the existence of all striking or anomalous drainage channels. An extreme example of this is shown in figure 2.6, where southwards from Lake Pickering are two further truly enormous bodies of water, 'Lake Humber' and 'Lake Fenland'. Another lake, 'Lake Harrison', may have occupied a large area of the West

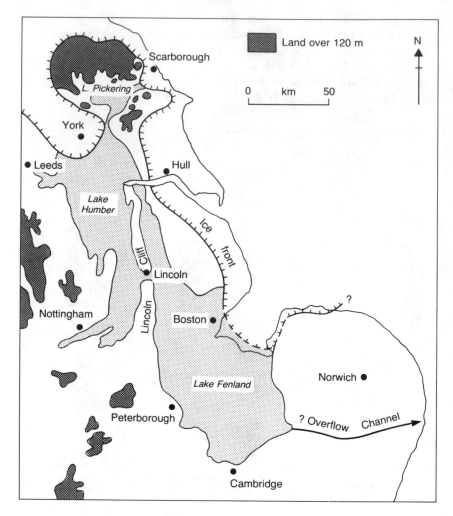

Figure 2.6 The possible limits of Pleistocene proglacial lakes in eastern England. If such lakes existed there is no reason why they should have existed simultaneously (after Whittow and Hardy, 1971, figure 17).

Midlands between Gloucestershire and Leicester (figure 2.7), while in north-western England and the Welsh Borderland 'Lake Lapworth' was postulated to have inundated an area stretching from north of Manchester to south of Shrewsbury (Poole and Whiteman, 1961). Features like the Thames's passage through the Chilterns via the Goring Gap and the Bristol Avon's gorge at

Clifton, were seen as the consequence of overflow from glacially dammed lakes, while the formation of Lake Lapworth was seen to explain the anomalous course of the River Severn (e.g. Lapworth and Watts, 1910, p. 768):

> The present Upper Severn, with its tributary, the Vyrnwy, was in pre-glacial times one of the headwaters of the Dee, and flowed northwards. Dammed back by the Northern and Welsh glaciers and ice sheets, the ponded waters overflowed at the lowest notch . . . at Ironbridge

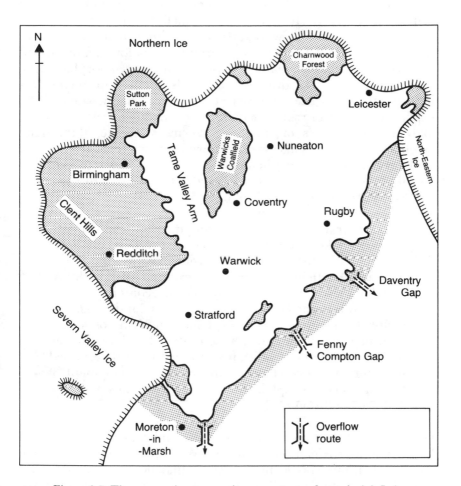

Figure 2.7 The approximate maximum extent of proglacial Lake Harrison in the English Midlands after the work of Shotton (1953).

. . . the north-western water, diverted across the Ironbridge overflow, began to cut the notch deeper until, in the course of time, it was deep enough to drain off from the Dee basin all the waters of the present Upper Severn and its tributaries, and in this way much of the pre-glacial upper Severn became transferred into the present Middle Severn . . . This theory accounts also for the anomalous course taken by the Severn and its remarkable swing from a north-east direction to one south-east and southerly.

There was an inevitable reaction against such grandiose mental constructions. In particular, Sissons (1960/61) demonstrated the limitations of Kendall's work, stressed the importance of stagnant ice and deglaciation in explaining certain 'anomalous' features, and referred to current observations of Scandinavian geomorphologists. He argued that odd drainage channels do not in themselves indicate the presence of lakes; that many channels could be explained by the presence of meltwater streams developed on, in and beneath masses of stagnant, downwasting ice; that many channels had features like up-and-down long profiles that were best explained as subglacial phenomena; and that concrete evidence for lakes themselves (e.g. shorelines, deltas, lake deposits) is rare. As Price (1973, p. 7) has written:

The general acceptance of the development of ice-dammed lakes and the cutting of overflow channels dominated British glacial geomorphology until the 1950s. This is a particularly interesting example of the almost circular reasoning adopted by glacial geomorphologists during this period. A particular set of landforms, the meltwater channels, were explained in terms of an ice dam blocking normal drainage routes. The limits of the ice cover and the condition of the ice were determined from the existence of the channels. The development of the channels was explained in terms of the limits of the ice and the condition of that ice.

Having established that the wholesale application of the glacial lake model was deleterious to an understanding of glacial drainage modification, one should not conclude that no large glacial lakes existed. For example, as Worsley (1975, p. 116) has said of 'Lake Lapworth', 'It is fair to say that while the former existence of the lake is not absolutely proven, very much more of the evidence is explained by accepting the lake hypothesis than by rejecting it.' Likewise, Lake Harrison (Shotton, 1953) may well have been a major influence on the geomorphology of the Midlands, creating a shoreline feature along its bounding escarpment to the east, and flowing through a series of gaps in that escarpment (e.g. at Fenny Compton). Possibly its most significant effect was on major river systems, for overflow from Lake Harrison led to the establishment of the Warwickshire Avon, a river which now flows in the opposite direction to the original preglacial drainage pattern. However, even Lake Harrison's existence has been the subject of some recent debate, and the possibility has been mooted that it was never just one great lake but a series of smaller water bodies, with its alleged shoreline at 400 feet (122 m) the result of structural control (see Harwood, 1988, for a discussion of the evidence).

Meltwater channels

The demise of an over-catholic interpretation of the importance of glacially dammed lakes as a cause of anomalous drainage channels led to a greater appreciation of the role of englacial and subglacial water in creating such features (Sissons 1960/61).

Good examples occur in the Cheviot hills, especially on their eastern side and their variety has been stressed by Clapperton (1971, p. 365):

> The features vary considerably in form and dimension. Many are small depressions only a metre or two wide and deep throughout their lengths; some begin in this way and then develop large proportions. Several channels are impressive features over twenty metres deep, with narrow, V-shaped cross profiles, and at least seven reach over thirty metres in maximum depth. Between these extremes many display various characteristics of form and size. They are cut either entirely in bedrock or in drift, or partly in drift and then into underlying bedrock . . . Seldom do the meltwater channels exist as isolated features on the hillsides; they are generally grouped in systems with tributaries, distributaries and abandoned

sections. Furthermore, the most impressive systems occupy pre-existing cols and valley-heads.

Further examples have now been mapped and described from other parts of England and Wales (table 2.3). Figure 2.8 displays the arrangement of channels from Cheshire, North Wales and Pembrokeshire.

Table 2.3 Select references to British meltwater channels

Location	Source
Lincolnshire Wolds	Straw (1961)
Cheviots	Clapperton (1971)
Conway Basin, N Wales	Fishwick (1977)
South Lake District	Vincent (1985)
North Staffordshire	Knowles (1985)
West Pennines	Johnson (1985)
Telford District	Hamblin (1986)
Rhyl and Denbigh, N Wales	Warren et al. (1984)
Penrith area	Arthurton and Wadge (1981)
Rhinog Mts, N Wales	Allen and Jackson (1985)
Vale of Eden	Burgess and Holliday (1979)

Not all meltwater channels have surface expression, sometimes being choked by later deposits, but they still provide an indication of the degree of erosion that can be achieved by subglacial waters. In East Anglia, for example, boreholes indicate the presence of deep, infilled channels lying well within the confines of present broad valleys. These 'buried tunnel valleys' (Woodland, 1970) have steep side-walls, which are sometimes vertical, irregular long profiles, and fills of clay, sands and gravels (Barker and Harker, 1984). The distribution of these large features in East Anglia is shown in figure 2.9. Their interpretation is not, however, without controversy for Cox (1985) has recently suggested that rather than

Figure 2.8 Examples of complex networks of glacial meltwater channels: (A) between Macclesfield and Congleton in Cheshire (after Johnson, 1985, figure 13.5); (B) in the Conway area of North Wales (after Fishwick, 1977, Figure 1); (C) in the Fishguard area of west Wales (after Sugden and John, 1976, figure 15.5).

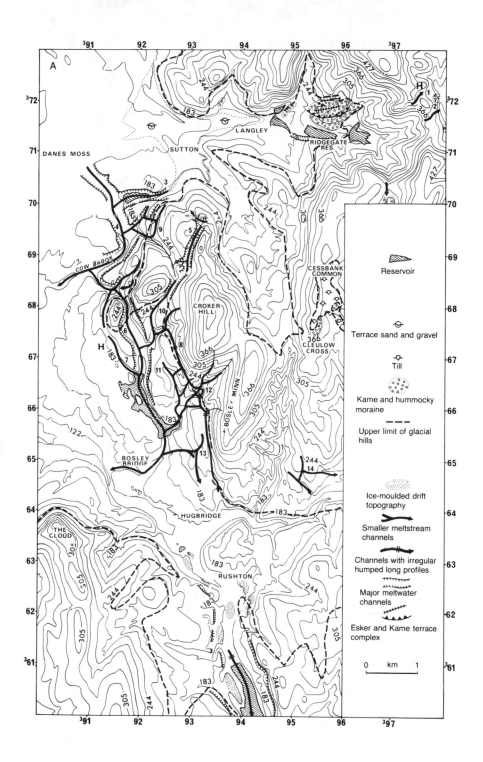

A

³91 92 93 94 95 96 ³97

DANES MOSS

LANGLEY

SUTTON

RIDGEGATE
RES.

COW BROOK

CESSBANK
COMMON

CROKER
HILL

CLEULOW
CROSS

H

BOSLEY MINN

BOSLEY
BRIDGE

HUGBRIDGE

THE
CLOUD

RUSHTON

Reservoir

Terrace sand and gravel

Till

Kame and hummocky
moraine

Upper limit of glacial
hills

Ice-moulded drift
topography

Smaller meltstream
channels

Channels with irregular
humped long profiles

Major meltwater
channels

Esker and Kame terrace
complex

0   km   1

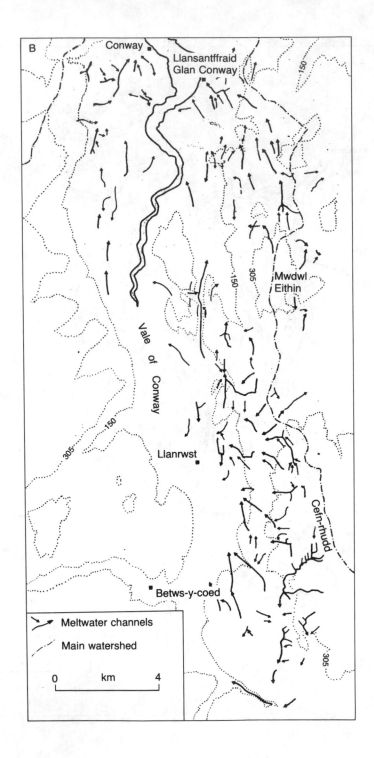

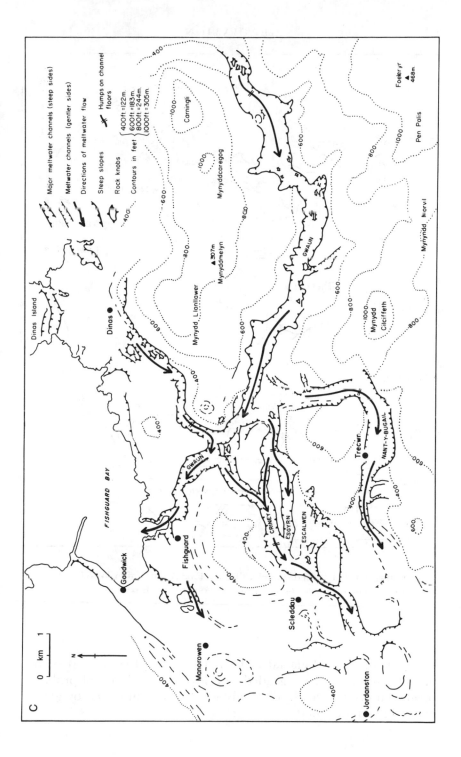

C

FISHGUARD BAY

Dinas Island

Goodwick

Fishguard

Manorowen

Scleddau

Jordanston

Dinas

GWAUN

GWAUN

CRINEY

ESGYRN

ESCALWEN

NANT-Y-BUGAIL

Trecwn

Mynydd, L.lanllawer

Mynyddcaregog

Carningli

▲307m
Mynyddmelyn

Mynydd
Cilciffeth

Pen Palis

Mynydd Morvil

Foeleryr
▲468m

Major meltwater channels (steep sides)

Meltwater channels (gentler sides)

Directions of meltwater flow

Steep slopes

Rock knobs

Contours in feet

400ft = 122m.
600ft = 183m.
800ft = 244m.
1000ft = 305m.

Humps on channel floors

400

600

800

1000

N

0   km   1

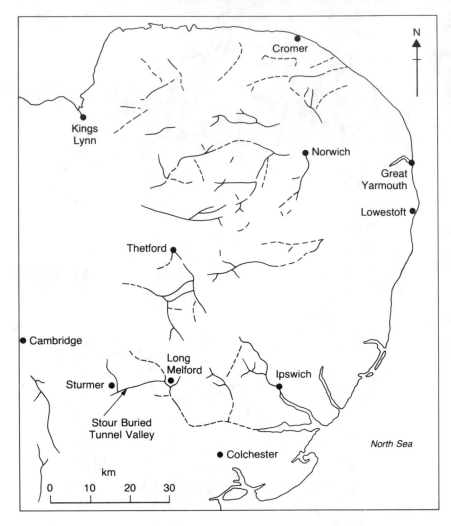

Figure 2.9 The buried tunnel-valleys of East Anglia (after Woodland, 1970). The origin of these features is controversial.

being overdeepened glacial valleys or subglacial features they are, in fact, ancient valleys, graded to a low sea-level and of probably preglacial origin, which have only locally been modified by glacial and subglacial agency.

The diversion of the Thames

The glaciation of England had dramatic effects on its major rivers. The courses of the Severn, Dee and Warwickshire Avon, for example, were greatly modified by the formation of glacial lakes and overflows (see p. 57). However, some of the most marked changes occurred in the course of the River Thames in its lower stretches.

The classic interpretation and description of these changes was made by Wooldridge (1938) and his co-workers, who postulated what has been called 'the double drainage diversion hypothesis'. The essence of this hypothesis was that in preglacial times following the retreat of the 'Calabrian Sea' (see p. 23) a proto-Thames flowed north-eastwards along the line of the Vale of St. Albans, and thence across central Essex to the North Sea. It was fed by rivers coming from the Weald across the present line of the Lower Thames. The ingress of a 'Chiltern drift' glaciation from the north west caused the first diversion of the proto-Thames, initially via the Borehamwood Channel and later to the Finchley Channel via two cols to the south of Watford. A second major ice advance, coming from the north and east and depositing the Chalky Boulder Clay, caused a second southward displacement by covering much of Essex and blocking off the Finchley Channel. This diverted the Thames more or less to its present route.

More recent investigations (summarized by Baker and Jones, 1980) have shown that while the proto-Thames routeway via the Vale of St Albans and a depression in central Essex can be substantiated, the subsequent evolutionary history was rather different from that postulated by Wooldridge. The 'double drainage diversion hypothesis' has now been replaced by a single, but somewhat complex, diversion created by Anglian ice. The envisaged sequence of events involves a series of advances, re-advances and retreats of the ice front, and a rejection of the Finchley depression as a routeway for the ancestral Thames and its reinterpretation as a proto-Wey–Mole valley (Gibbard, 1979).

Prior to the southward diversion of the Thames to its present course, the fluvial geomorphology of south-eastern England was markedly different to what we see today. For example, the River Medway took a north-easterly course across Essex, and deposited large quantities of gravels, which contain lithologies derived from the Weald (Bridgland, 1988). In other words, the shifts in the position of the Thames transformed eastern Essex from being, in effect, part of the northern Weald to being the northern flank of the Thames Valley.

Glacial deposition

Large areas of England and Wales are smothered in glacial deposits of various types (figure 2.10). The origin of such 'drift' was attributed, until the 1840s, to the activities of the Noachian Deluge, but from then on the importance of glacial agency was put forward. One of the first exponents of the glacial theory in England was a former exponent of the power of the Flood, Buckland (1840/41), who, having visited Switzerland and Scotland in the company of Louis Agassiz, the great exponent of the concept of *Die Eiszeit* (the Ice Age), was converted to glacialism and started to identify moraines in northern England (p. 346):

> Immediately below the vomitories of the eastern valleys of the Cheviots, enormous moraines are stated to cover a tract of four miles from north to south and four from west to east.

> In the vicinity of Penrith, near the junction of the Eden with the Eament and the Lowther, are extensive moraines loaded with enormous blocks of porphyry and slate brought down, by glaciers, which descended from the high valleys on the east flanks of Helvellyn, and in the mountains around Patterdale, into the Lake of Ullswater (considered to be then occupied by ice) and from the valleys by which the tributaries of the Lowther descend from the east flank of Martindale, from Haweswater and Mardale.

By convention, a distinction is normally drawn between those deposits laid down in the Devensian Glaciation (the so-called 'Newer Drift') and those laid down in earlier glacial phases (the 'Older Drift'). The former types of deposit have the clearest morphologies, and it is possible to identify relatively fresh landforms, be they kames, eskers, kettles, drumlins or moraines. By contrast, the older drift terrains to the south of the Devensian limit

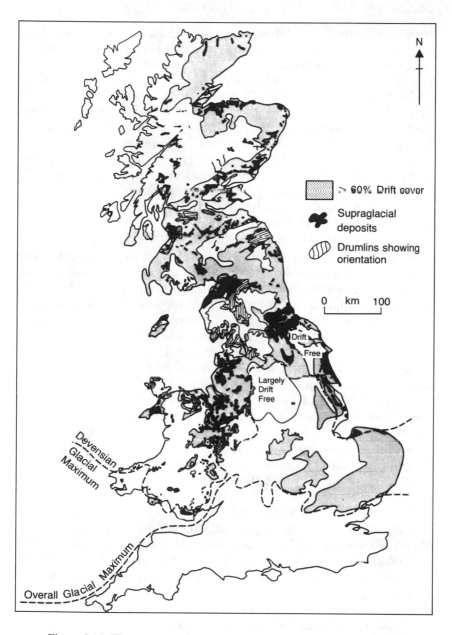

Figure 2.10 The extent of glacial deposits over Britain, and the distribution of drumlins (showing mean orientation) and supraglacial sediments within the late Devensian ice-sheet limit (after Boulton et al. 1977, figure 17.4).

have been so remoulded by subsequent events, especially periglacial activity, that the former presence of ice is primarily indicated by somewhat amorphous spreads of till and outwash gravel. In the case of some of the very oldest deposits (e.g. the Plateau Drift of the Oxford Region, and the Chiltern Drift) the degree of post-depositional modification has been such that it even becomes problematical to stipulate that their origin is unequivocally glacial. Wright (1937, p. 75) also identified profound differences in the drainage systems on the two types: 'Extensive denudation and maturity of drainage which characterises the older drift is suggestive of a very considerable lapse of time since its exposure to subaerial erosion. On the other hand the post-glacial denudation of the more northerly drift is quite trivial.'

However, as Wright himself recognized (1937, p. 76) such a two-part classification is in reality oversimple and both the older and newer Drifts are capable of further subdivision. A discussion of the full complexities of Pleistocene glacial stratigraphy is, however, beyond the scope of this volume, for, as we saw at the beginning of this chapter, it is a matter of great complexity and controversy.

It is also possible to subdivide till deposits according to their area of derivation. So, for example, in northern England ice came from a variety of centres. Local ice sheets developed in the Lake District and in the northern Pennines; other ice came from the Southern Uplands of Scotland and from Scandinavia. The identification of the tills resulting from these different sources is facilitated by the presence of many characteristic erratics: Eskdale granite and Ennerdale granophyre in the tills derived from Lake District ice; Criffel granite and Cheviot igneous rocks in tills derived from the Southern Uplands ice; and Norwegian erratics in the Scandinavian drift.

Likewise, in the Midlands, ice advanced from three main sources, each resulting till type having its own erratic suite: the Welsh till contained Uriconian volcanic rocks; the Eastern drift contained Mesozoic rocks; the Irish Sea drift carried sediments from the Irish Sea, together with Ailsa Craig microgranite and Goat Fell granite from Arran.

Long-travelled erratics are also found in South Wales. Here too one finds rocks from Ailsa Craig and Arran, volcanic rocks from North Wales and Cretaceous flints from Northern Ireland or the bed of the Irish Sea.

Drumlins

Drumlins are generally regarded as elongated hills with a rounded blunter end towards the ice and a longer, tapering end in the down-ice direction. These ideal forms approximate to a half-egg shape, the egg being sliced through in the plane of its longest axis. Thus fields of drumlins have often been termed 'basket of eggs' topography (Sparks and West, 1972, p. 72). However, in reality, the definition of drumlins is problematical for they may range from small circular mounds to elongated rib-like features and there may be a hierarchy of forms with small drumlins superimposed on large streamlined forms called megadrumlins (Riley, 1987). Furthermore, they may vary from pure drift forms developed in subglacial till, through features with a core of rock, to streamlined aerofoil type forms moulded out of relatively non-resistant rock. Characteristically, drumlins are between 150 and 1,000 m in length and 75 to 500 m in width.

Bearing such problems of definition in mind, it is none the less possible to see a relatively clear pattern in the distribution of British drumlin features (figure 2.11), all of which occur within the limits of the New (Devensian) Drift. One of the most extensive drumlin fields is that of north-west England, most notably in the Eden Valley and Solway (figure 2.12) – a belt that continues into south-west Scotland and across the Tyne Gap into north-east England. It almost links up with another field to the south of the Lake District (around Ulverston, Barrow-in-Furness and Kendal). To the west of the Lake District the drumlin belt swings round towards Maryport. Other groups occur in the Ribble Valley and in Wensleydale. Thus the uplands of the Lake District and North Pennines are almost entirely surrounded by drumlins, although they are restricted to the lower ground, only rarely occurring above 300 m (King, 1976; Embleton and King, 1967). They also tend to be concentrated at the foot of steeper slopes and on cols and divides, suggesting to King (1976, p. 118), 'that they tend to occur in those areas where the movement of the ice was constrained partly by reduction of gradient and partly as a result of interference by other ice masses or glacial diffluence.' In such situations ice thickness or speed would be increased, creating greater pressures on the glacier bed and, where suitable material is available, this could well be shaped into drumlin form.

Other major drumlin fields are on the Isle of Man, in Anglesey

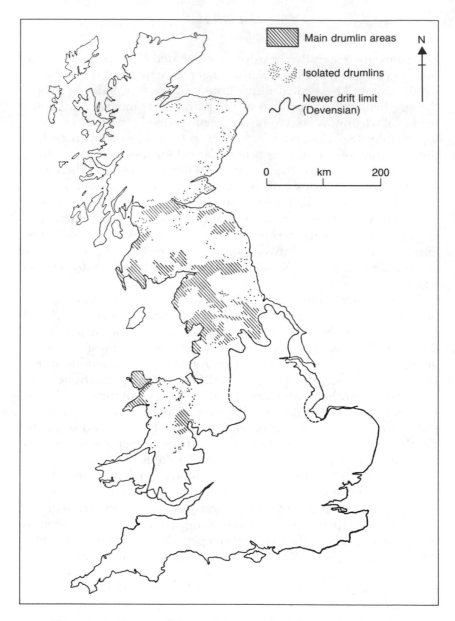

Figure 2.11 The distribution of drumlins in Britain (modified after Embleton and King, 1967, figure 14.4).

and in the Welsh border country near Wrexham, Rhyl and Denbigh (Warren et al., 1984) and on the borders of the Cheviots (Clapperton, 1971).

The distribution of drumlins is related to areas of deposition of thick subglacial till which have not been subsequently covered by

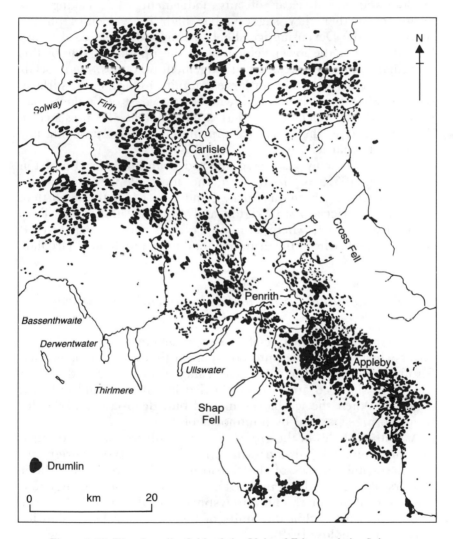

Figure 2.12 The drumlin field of the Vale of Eden and the Solway lowlands (after the work of Hollingworth, in Embleton, 1984, figure 7.22).

supraglacial deposits or proglacial outwash. The distribution of thick subglacial till may in turn depend on glacier sliding velocity. Boulton et al. (1977, p. 243) suggest that, 'where sliding velocities do not exceed the order of 50–100 m/year it is found that there is a partial or complete lodgement till cover on the subglacial bed, whereas at higher sliding velocities the subglacial bed is actively eroded and swept clear of any lodgement till.' Taking this argument further they suggest that high ice velocities on the margins of the Devensian ice sheet in England inhibited drumlin formation; that the lower velocities on the up-glacier side of this marginal zone may have allowed lodgement till deposition to occur for a sufficiently long time for large drumlins to form on its surface; that the absence of drumlins down the east coast of England may reflect high ice velocities in that area, whereas the presence of drumlins near to the Welsh ice-cap margin reflects probable low velocities in that ice cap.

Ice velocities may also have affected drumlin form (Embleton and King, 1967, p. 336), with the most elongated varieties occurring where the ice was thickest and fastest (e.g. to the north of the Lake District around Wigton and Aspatria, and in the Tyne Gap). By contrast, where a basal ice shed existed (e.g. around Appleby and Carlisle) the drumlins tend to be much more circular.

Ice stagnation basins and supraglacial landsystems

The stagnation and wasting of ice sheets can create a series of basin forms which may persist into post-glacial times. One type of such feature is the kettle hole, formed at a stagnant ice front as detached ice blocks slumped and became buried by ablation and outwash deposits. When the ice blocks melted out, deep crater-like holes would form, separated by hummocks of drift.

As with all holes in the ground it is difficult to establish origins, but one group of depressions which may have this type of origin are the meres and mosses of the Shropshire–Cheshire Plain. Major clusters are centred around Delamere, Knutsford, Congleton, Ellesmere, Whitchurch and Shrewsbury, and the great bulk occur in drift deposits within the limits of the main late Devensian ice advance (Reynolds, 1979).

Another extensive area of hollows was represented by the meres of East Yorkshire, although most of them have now been drained and only Hornsea Mere survives. Some of them were kettle holes, some resulted from uneven deposition of till, and some of them are sites of former valleys in the underlying chalk (Flenley, 1987).

Supraglacial landsystems generally develop when sediment is deposited on the surface of a glacier. The sediment is for the most part subglacial in origin, and is exposed on the glacier surface as a result of the pattern of ice flow. As a consequence of this supraglacial deposition, bodies of dead glacier ice become buried within the sedimentary sequence, giving an irregular pattern of landforms and sediments (Paul, 1983). Compressive flow is required for supraglacial deposition to occur and commonly results where a large bedrock obstruction, such as an escarpment, causes the ice to decelerate. Compression may also occur between ice lobes or at the outer limits of glaciation when the ice margin is slow moving, and dams more active ice up glacier.

The largest area of supraglacial deposition in Britain occurs on the west flank of the Pennines in Cheshire and the northern flank of the Staffordshire coalfield. It formed when Irish Sea ice impinged on the west Pennine escarpment in the Devensian and produced a sweeping belt of hummocky topography between Macclesfield and Whitchurch (Thomas, 1989). Another significant area of supra-glacial topography occurs in the Vale of York.

Outwash phenomena

The great Pleistocene ice sheets would have liberated large amounts of meltwater that would have transported and deposited huge amounts of sediment. Curiously, however, whereas in the north European lowland evidence for meltwater activity is very common, the same cannot be said for lowland England (Sparks and West, 1972, p. 84). There are, however, exceptions from North Norfolk (figure 2.13a) (Sparks and West, 1964), where south-eastwards from Blakeney there are two clear outwash plains (the Kelling Ice Front and the Salthouse Ice Front) together with a whole series of Kames, Kame terraces, and a sinuous ridge known as the Blakeney 'esker', the origin of which is controversial (Gray, 1988). This group of features, though remarkably fresh in form, is attributed to the penultimate glaciation. Further west, near Hunstanton, there are various Devensian outwash features including the Heacham Kame Deposits and the Hunstanton Esker (figure

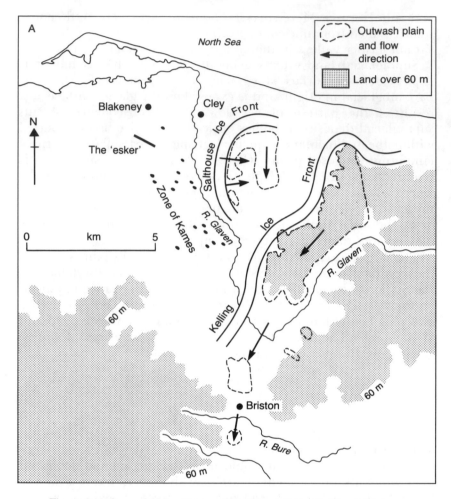

Figure 2.13 Some drift landforms of Norfolk: (A) the Kelling-Salthouse complex; (B, facing page) the Heacham area to the south-east of Hunstanton (modified after Sparks and West, 1964 and Boulton, 1977, figure 22).

2.13b). The latter is probably a true englacial feature (Boulton, 1977).

Conclusion

Although the glaciation of England and Wales only occurred sporadically during a period after 2.4 million years ago, the

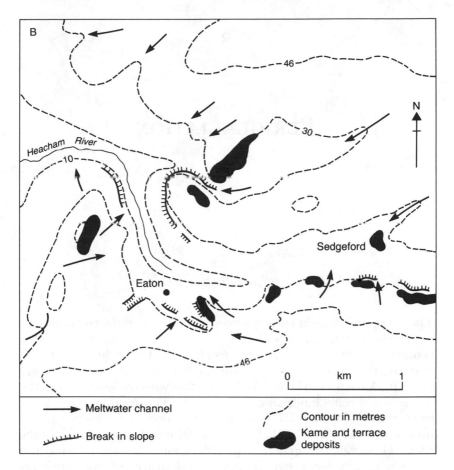

B

Heacham River

Sedgeford

Eaton

N

46

30

10

46

0   km   1

→ Meltwater channel

↙↙↙↙↙ Break in slope

⌒⌒ Contour in metres

Kame and terrace deposits

glacial impact on the landscape north of a line from Bristol to London was out of all proportion to the length of time involved. The three main processes involved in this impact were the erosion of highlands and lowlands, the deposition of large quantities of till, and the release of large volumes of meltwater.

However, the landforms of Britain were also substantially modified in another way during the cold phases of the Pleistocene, and that is by a group of processes to which the name 'periglaciation' is often given. That is the subject of chapter 3.

3

PERIGLACIATION

Introduction

The geomorphological consequences of intense periglacial conditions in the Quaternary have been recognized in Britain for more than a century. Waters (1978) cites the work of De La Beche (1839) who applied the term 'head' to superficial deposits in Cornwall; Godwin-Austen (1851) and Fisher (1866) who recognized that the 'head' was a relict landscape feature and was caused by Pleistocene frost action; Wood (1882) who understood the significance of involutions; and Reid (1887) who explained the origin of coombe rock. However, in general, British geomorphologists did not become greatly interested in periglacial processes and products until the 1960s, and Waters attributes the stimulation of periglacial geomorphology at that time to Peltier's (1950) paper on 'the geographic cycle in periglacial regions'; to contact with continental geomorphologists who in 1954 produced a journal devoted to periglacial studies, the *Biuletyn Peryglacjalny*; and to the visit to Britain from the Antipodes of Te Punga (Te Punga, 1957). Few would now doubt the importance of periglacial processes and forms in the British landscape. This is especially true of the Chalk of southern England: much of it, lying outside the maximum limits of glaciation, was exposed to prolonged tundra conditions; because of the lack of contemporary fluvial activity Pleistocene forms have been preserved; and Chalk is highly susceptible to frost attack (Williams, 1986).

Solifluction spreads

Solifluction spreads have been identified on hill-slopes throughout much of Britain and it is clear that colluvium of a soliflual origin is one of the most widespread and morphologically important surface sediment types.

Of particular fame are the extensive sheets of 'coombe rock', composed of soliflucted Chalk and flint, that mantles the foot-slopes beneath most Chalk escarpments and which emerge as large fans at the mouths of the larger scarp face dry valleys. Dip slope valley systems may also contain considerable depths of chalk 'head' and cliff sections on the Chalk coasts of Dorset, Kent and East Sussex reveal sections through truncated, hanging dry valleys which contain coombe rock sludge resting upon frost-shattered Chalk. The greatest known thickness of such accumulation occurs in the section at Black Rock, Brighton, where 20 m of superficial material overlies a raised beach (Jones, 1981, p. 236).

Although possibly most easily recognized on the Chalk (e.g. Horton et al., 1981), solifluction spreads are known from other lithologies in south-east England. For example, Skempton and Weeks (1976) describe deposits south of the Greensand escarpment in the vicinity of Sevenoaks. There, small hills and ridges, standing up to 20–30 m above the streams of the Weald Clay Vale, are capped by deposits consisting of chert fragments (up to 20 cm in length) and other stones derived from the Greensand, set in a clay matrix. These deposits are interpreted as the dissected remnants of originally more or less continuous sheets which spread off the scarp and moved down the gentle slopes of the then-existing land surface for distances of at least 2 km (and sometimes as far as 4 km) on slopes of less than 2° (and sometimes as low as 0.8°). Skempton and Weeks invoke intense frost shattering and periglacial solifluction (probably in the Wolstonian) to explain the release of material from the Greensand and the movement of such large materials in such large quantities over such gentle slopes.

Another lithology that has produced extensive forms of ill-sorted periglacial material of soliflual and torrent origin is the Jurassic escarpment of the Cotswolds (Tomlinson, 1941).

Solifluction has proved to be capable of moving very large calibres of material, as is made evident by the far-travelled clitter blocks in the valleys and on the slopes around Dartmoor and Bodmin Moor. The most spectacular instances are the block-

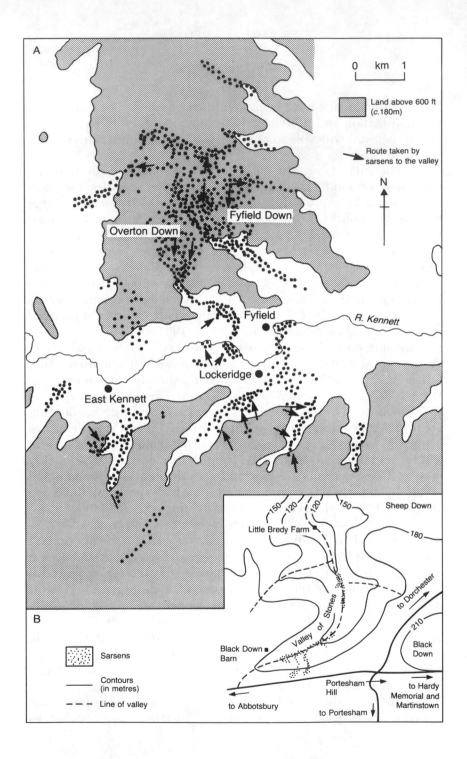

A

0 km 1

Land above 600 ft (c.180m)

Route taken by sarsens to the valley

N

Fyfield Down

Overton Down

Fyfield

R. Kennett

Lockeridge

East Kennett

B

Sarsens

Contours (in metres)

Line of valley

Sheep Down

150 120 120 150

Little Bredy Farm

180

to Dorchester

Valley of Stones

Black Down Barn

210

Black Down

Portesham Hill

to Hardy Memorial and Martinstown

to Abbotsbury

to Portesham

streams composed of sarsen blocks that occur on the chalklands of Dorset, Wiltshire and Berkshire. The sarsens themselves, sometimes called 'greywethers', are quartzitic sandstones, flint breccias and conglomerates (pudding stones) of Tertiary age (Summerfield and Goudie, 1981). They occur as blocks up to 200 tons in weight, and 4–5 m long. The blocks are generally tabular in shape, with a thickness up to 2 m, and give the impression of being slabs that are the remnants of a formerly jointed and more continuous stratum. These huge slabs have been transported from plateau tops, down small, gentle dry valleys, and into through-flowing river valleys. Good examples occur in the Marlborough Downs and Kennett Valley, (figure 3.1A) and near Portesham in Dorset (figure 3.1B). In the Marlborough area some of the blocks have moved as much as 4,000 m on an average gradient of between 1 and 3° (Small et al., 1970).

The significance and origin of 'head' has been understood for over a century (see Dines et al., 1940 for a review of early work). Godwin-Austen (1851, p. 123) described 'head' from various locations:

> The thickness which these accumulations attain is frequently very great; the mass described from Borlase varied from 50 to 20 feet. West of Pendennis it may be estimated at nearly 30 feet, and reaches to 20 feet in the district of the Prawle. In the coast sections of the Channel Isles, I have measured it from a few feet to as much as 40. And everywhere about the shores of the western entrance into the Channel it is of sufficient thickness to constitute the line of low under-cliffs.

Some time later, Wood (1882, p. 718) recognized the importance of frost action and permafrost in creating head and, although the term had yet to be coined, described the process of solifluction:

> In Arctic countries, like Siberia, the soil is permanently frozen to a considerable depth, only the upper two or three feet of it thawing during summer . . . It is obvious, however, that the part below remaining frozen is impermeable by

Figure 3.1 Sarsen blockstreams in southern England: (A) in the Marlborough Downs area of Wiltshire; (B) in the Portesham area of Dorset

water, so that, as this can have no escape vertically, it must convert the thawed layer into sludge; the tendency of which is to slide horizontally from higher to lower ground, thus accumulating more and more in depressions, and exposing more and more of the surface from which it slid to the operation of this agency, and so keeping up the supply. The frost and thaw then acting upon limestone-rock, which is partially porous, has split it up into angular fragments, which become dispersed in the resulting sludge. . .

Many coastal sections in western Britain reveal extensive head deposits overlying raised beaches and banked up in front of fossil cliff-lines (Mottershead, 1971). Detailed descriptions are given for the Aberystwyth area in Wales by Watson and Watson (1967). Equally, many of the valleys of south and central Wales have ramp-

Plate 3. The Valley of Stones (or Valley of Rocks) in North Devon displays scree slopes, angular tors, and a deep fill of solifiucted material. There is some argument whether this valley is a relict glacial overflow channel or a subaerial valley truncated by cliff retreat. (Author)

like solifluction terraces running down to the modern streams (Crampton and Taylor, 1967), while sections in the Valley of Rocks in North Devon have indicated that locally there is a 35 m fill of soliflucted material (Dalzell and Durrance, 1980).

There are remarkably few precise dates for solifluction spreads in western Britain. An exception is provided by Scourse's work on solifluction breccias in the Scilly Isles (Scourse, 1987) where 29 ^{14}C dates within the lower breccia indicate that deposition occurred between 34,000 and 21,000 years BP (Middle and Late Devensian). The colluvial processes that produced the breccia also caused the exhumation of the Scilly Isle Tors at that time.

Large solifluction sheets, probably of Loch Lomond Stadial age, occur in the Cheviot Hills, and consist of features characterized by bluffs 3–20 m high and smooth treads 20–300 m wide. They are composed of three types of material: soliflucted till; soliflucted growan (regolith) and soliflucted frost shattered material (Douglas and Harrison, 1987).

Patterned ground and periglacial structures

One of the most prominent features of present-day tundra areas is the presence of various types of patterned ground. In some upland areas of Britain some small-scale features are undoubtedly still forming today (Hollingworth, 1934; Pemberton, 1980), but in the past the features were both much larger and more widely distributed. The patterns are predominantly composed of various types of polygons and stripes (Evans, 1976). Excellent examples are reported from such areas as the Lake District, Snowdonia, the Stiperstones of Shropshire, the Pennines, the Midlands and East Anglia (table 3.1). Indeed, the maps of fossil ice-wedge distribution (e.g. Williams, 1969) indicates very clearly that at various points in the Pleistocene permafrost was widespread even at low altitudes (figure 3.2). Williams (1969, p. 408) was especially struck by their persistence in eastern England and the Midlands:

> . . . the abundance of structures suggests that continuous permafrost must have existed in certain districts. Perma-frost structures underlie an estimated 50 per cent of the land surface in such areas as the lower Thames valley and the Breckland of western Norfolk and Suffolk. Permafrost

Table 3.1 Select references on patterned ground phenomena

Location	Source
Ullswater, Lake District	Hay (1937)
Lake District	Caine (1972), Hollingworth (1934)
English Midlands (Wolverhampton)	Morgan (1971)
Rhinog Mountains, N Wales	Foster (1970)
Snowdonia, N Wales	Ball and Goodier (1970)
Stiperstones, Shropshire	Goudie and Piggott (1981)
Pennines	Tufnell (1969)
Severn Valley, Gloucestershire	Allen (1984)
Wiltshire and Breckland	Evans (1976)
Hampshire	Lewin (1966)
Cambridge	Paterson (1940)
Devon	Jenkins and Vincent (1981)
Oxford Region	Seddon and Holyoak (1985)

was probably continuous, since not every place where it occurred would be occupied by structures, and not all structures would be preserved.

The density of structures is, however, seemingly rather less in the south-west of England (figure 3.3), although fine examples do occur at higher altitudes on Dartmoor where the granite clitter radiating from tors (e.g. Middle Staple Tor near Merrivale) is arranged in spectacular stripes, festoons and circles (Te Punga, 1957). There are also good examples in apparently highly susceptible ball clays at lower altitudes near Bovey Tracey (Jenkins and Vincent, 1981). According to Williams (1969, p. 407):

> The best explanation for the scarcity of permafrost structures in the southwest is that mean annual temperatures were too high. Invasion of Atlantic air moderated the climate sufficiently to prevent extreme winter cold. However, the mild maritime air does not seem to have extended its influence far across the country. Extensive permafrost seems to have occurred as far west as West Sussex and Salisbury Plain.

The presence of so much permafrost in England and Wales during the Pleistocene cold phases indicates, by analogy with modern

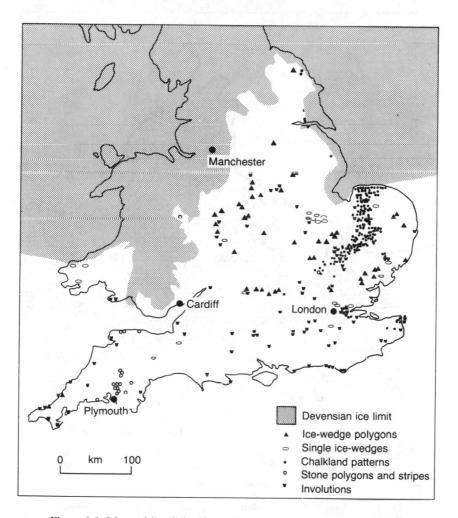

Figure 3.2 Map of localities for patterned ground phenomena presumed to have formed during the Devensian glaciation (from
Williams, 1969, figure 1, p. 401).

areas underlain by permafrost, that climatic conditions were
indeed extreme. Williams (1975) estimates that over large areas
mean annual temperatures would have been − 8 to − 10°C or less.

Patterned ground phenomena have many different forms. The
following are some examples of the types that have been described.

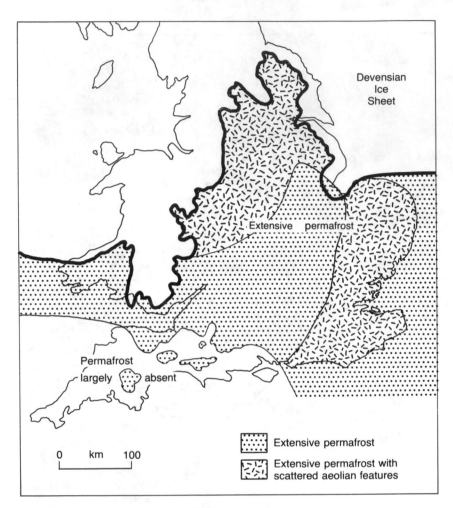

Figure 3.3 Conjectural distribution of permafrost in England during the Devensian. The scattered aeolian features include loess, dunes and ventifacts (from Williams, 1969, figure 2, p. 406).

CHALKLAND POLYGONS AND STRIPES

In the Breckland of East Anglia, at locations such as Thetford Chase and Grimes Graves (figure 3.4) there are a series of well-developed stripes (on slopes) and polygons (on low-angle surfaces). They are represented at the surface by vegetation, with alternations of grass and heather (*Calluna*). This reflects a pattern of subsurface

Plate 4. Vegetation stripes on Thetford Chase in the Breckland. The dark stripes are composed of heather (*Calluna*) and the lighter stripes are composed of grasses. (Author)

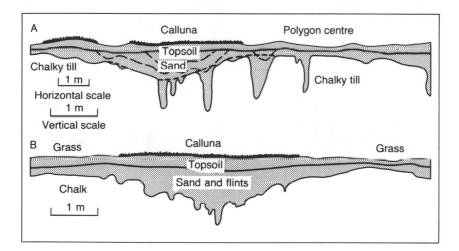

Figure 3.4 Cross-sections through Breckland polygons and stripes (after Williams, 1964).

sediment conditions, with the heather occurring above deeper pockets of sandy drift and the grass occurring where the Chalk or Chalky Till comes closer to the ground surface. The average distance between stripes is approximately 7.5 m and that between polygon centres about 10 m.

ROCK POLYGONS AND STRIPES

Around the stiperstones of Shropshire (figure 3.5) coarse angular boulders (up to 2.5 m long axis length) form 3 m wide stripes which extend for 200–300 m over slopes with angles of 6 to 16°. On lower angle slopes there are irregular polygons with diameters of around 7–9 m (Goudie and Piggott, 1981).

Plate 5. Intense frost attack on the Ordovician Arenig quartzites of the Stiperstones in Shropshire has produced angular debris that has been sorted into a series of stripes and polygons. (Author)

Figure 3.5 The distribution of the stripes between Cranberry Rock and the Devil's Chair, Stiperstones, Shropshire (after Goudie and Piggott, 1981, figure 4).

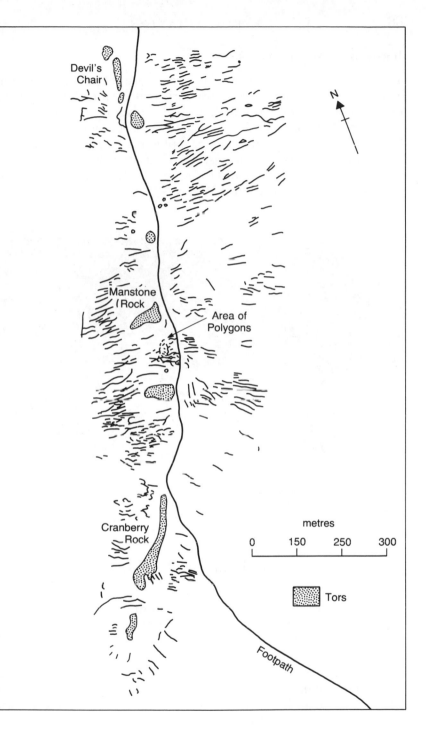

Devil's
Chair

N

Manstone
Rock

Area of
Polygons

Cranberry
Rock

metres

0   150   250   300

Tors

Footpath

ICE-WEDGE PSEUDOMORPHS

These are perhaps the most widespread form of patterned ground. The wedges themselves are characteristically 0.5 to 3.0 m deep and up to about 1.5 m across. They form polygons ('palaeomosaics' and 'tessellations') (Worsley, 1966) which sometimes have a hierarchy of forms, with master polygons 8–12 m across, and subsidiary ones 3–7 m across. Good descriptions are given by Shotton (1960) and Morgan (1971) for the English Midlands and by Watson (1981) for central Wales.

Plate 6. An exposure of angular 'head' from the northern coastal fringe of Exmoor showing an ice wedge feature and the angular and confused nature of the material. (Author)

Pingos and related forms

A number of types of mound are produced by the growth of segregated ice in the ground. The mounds are due to the growth of ice lenses, or 'hydrolaccoliths', which lift the ground above them as they grow. Some of the mounds may be only a few metres across, while others, called pingos, may be as much as 300 m in diameter and 60 m high. Pingos, which today occur in areas like the Mackenzie Delta in north-west Canada, are intimately associated

with the presence of permafrost. Two main types exist: the closed
system and the open system. Closed-system pingos result when a
lake is drained and infilled and the unfrozen ground, known as
talik, beneath the lake is progressively frozen from all sides. The
resultant increase in volume is taken up by the growth of an ice lens
which lifts the ground surface. The more common open-system
pingo occurs when the ice lens is nourished by extraneous water,
flowing, for example, from an aquifer. With both types of pingo, as
the mound develops, the uplifted soil layer is gradually eroded
away to expose the ice core, which then melts (figure 3.6). The
effect of this transference of material from the mound to its margins
is to produce, when melting occurs, a rampart of sediment around a
central depression.

The remnants of such ramparts and depressions have been
identified in various parts of Britain (figure 3.7), and can be used to
infer the existence of permafrost. Probably the first correct
identification was by Pissart (1963) of examples near Llangurig,
mid-Wales. A notable group occurs in south-west Wales between
Aberystwyth and Cardigan (Watson and Watson, 1974) (figure
3.8) and were probably of the open-system type. In the Chedlyn
Valley of northern Dyfed the largest example has an external
diameter of around 200 m and a rampart 6 m high (Bowen, 1977).
Handa and Moore (1976) attribute the mid-Wales pingos to Loch
Lomond advance permafrost conditions.

Pingo scars have also been located in the Whicham Valley of
Cumbria. For example, Grass Guards Pingo Scar at SO 158844 is
an almost circular hollow of around 50 m diameter, with a hollow
that is 4 m deep from the rampart crest to the base of the organic
fill. There are also some depressions with raised rims in the
Shropshire–Cheshire plain (Reynolds, 1979), in Surrey (Bryant
and Carpenter, 1987) and in central London (Hutchinson, 1980a).
However, some of the most striking rampart and depression
complexes occur in East Anglia. For example, 3 km south-east of
Conington in Cambridgeshire, on the margins of the Fens, are two
elliptical depressions up to 1,000 m long, which are formed
in Oxford Clay and superficial drifts, and which have raised
ramparts (Burton, 1976). A major complex of depressions (generally
10–120 m in diameter) occurs on Walton Common in Norfolk and
elsewhere on areas of lowlying Chalk between the boulder clay
uplands of East Anglia to the east and the Fens to the west (Sparks
et al., 1972).

Some of these features may not necessarily result from pingo

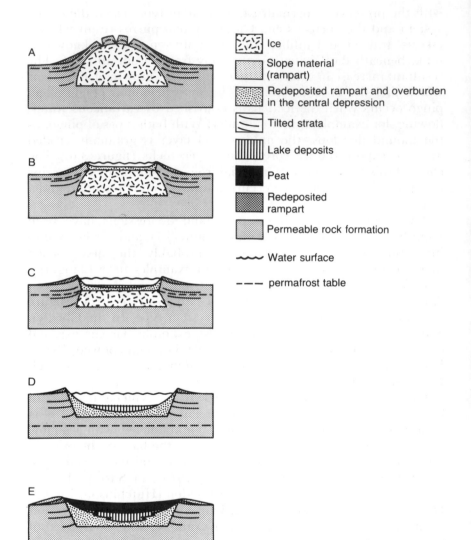

Figure 3.6 A five-stage model of pingo decay, showing the development of the central depression and the rampart. Stages C and D represent modern pingo scars, while E shows an ancient pingo scar (after de Gans, 1988, figure 13.3).

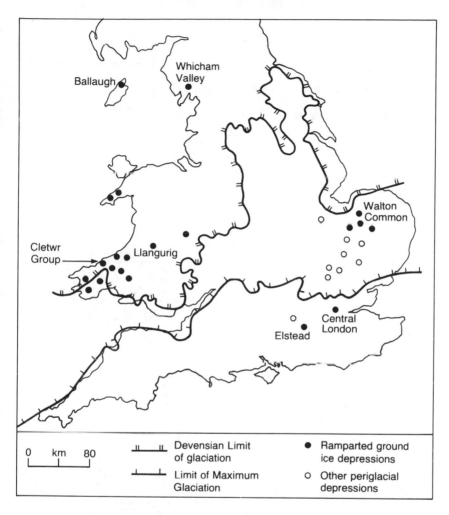

Figure 3.7 Generalized distribution of areas of ramparted ground ice depressions and other periglacial depressions (modified from Bryant and Carpenter, 1987, figure 16.1).

growth but from other types of ground-ice segregation and may best be described by some cautious name as 'presumed ground-ice depressions' (Sparks et al., 1972) or 'ramparted ground ice depressions' (Bryant and Carpenter, 1987).

Some features may be remnants of palsas – ice cored mounds, often with a peat cover, which formed without the benefit of

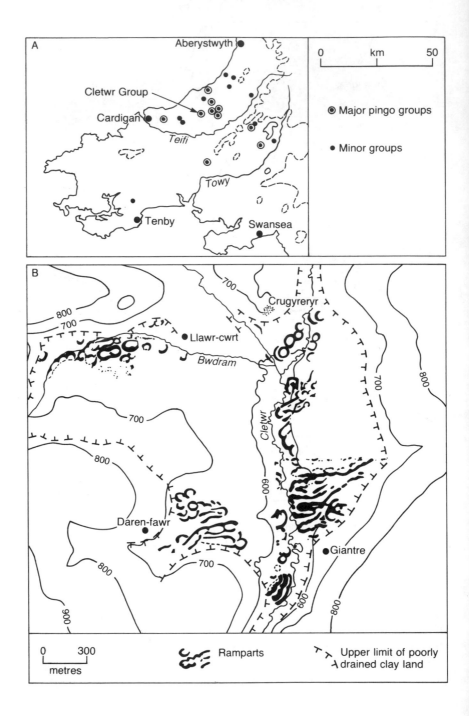

A

Aberystwyth

Cletwr Group

Cardigan
Teifi

Towy

Tenby

Swansea

0   km   50

◉ Major pingo groups

● Minor groups

B

700

Crugyreryr

800
700

Llawr-cwrt

Bwdram

Cletwr

700

700

800

600

Daren-fawr

Giantre

800

700

600

900

800

0   300
metres

Ramparts

Upper limit of poorly
drained clay land

hydraulic injection of the type required to form a pingo. Other terrain irregularities may be a consequence of the burial and melt of former icings (*naleds*) on aggrading floodplains (Coxon, 1978). Catt et al. (1982) postulate the existence of such features on the floor of a chalk dry valley in Wessex. They suggest that the icings arose through the winter freezing of springs which discharged on to the valley floor. During the summer thaw period snowmelt-derived run-off buried the ice bodies beneath fluvial gravels. Following burial the ice eventually melted and induced collapse of the overlying gravels to produce a series of irregular hollows.

Stratified screes

Screes, including the widespread examples for Wales dating to the Loch Lomond Advance (Ball, 1966), are accumulations of coarse, angular rock debris derived from high angle cliffs or rock faces, mainly by frost-splitting and subsequent rock falls. Further movement may occur by creep, sliding or rolling. Most screes display limited stratification or preferred orientation of constituent clasts. However, in the case of fine screes formed beneath slopes of appropriate lithology (e.g. slate, shale, mudstone) which are easily frost-shattered into flattish fragments, the coarser clasts are orientated parallel to the slope and interbedded with silty, less stony layers. Such deposits, which show dips of up to 30°, are called stratified screes or *grèzes litées* (Boardman, 1978).

Stratified screes are believed to have attained their distinctive characteristics because of seasonal alternations between frost-shattering and reworking of the fallen rock fragments by sliding, creep and sheetwash. Water movement also appears to have played a role in their formation, and it is possible that underlying permafrost and the presence of snow patches could contribute to this (Boardman, 1985). They are extensively developed in the shale terrains of central Wales and show a markedly preferred orientation with respect to aspect, with over 75–80 per cent of the screes having an aspect between 140 and 270° (Potts, 1971). The explanation for this is that the number of freeze–thaw cycles to produce debris

Figure 3.8 Pingo remains in south-west Wales: (A) distribution; (B) pingo forms in the Cletwr basin. The contours are in feet (modified from Watson and Watson, 1974, figures 1 and 2).

might be greater on west- and south-facing slopes. Other examples
of stratified screes occur on mudstones on the slopes of Latrigg near
Keswick (Boardman, 1985).

Protalus ramparts

Another type of cryonival feature, caused by scree formation, is the
protalus rampart. These are arcuate accumulations of angular
talus that occur at some distance from the parent bedrock cliff. The
ramparts are separated from this cliff by a depression.

This particular form arises (see figure 3.9) where rock blocks,
released from the cliff by frost-shattering and rock fall, have fallen
or slid across a perennial snowpatch. They are thus sometimes
called 'snow-slope foot accumulations'.

Protalus ramparts are known from the Lake District at altitudes
between 300 and 600 m OD; they have aspects in the north-east

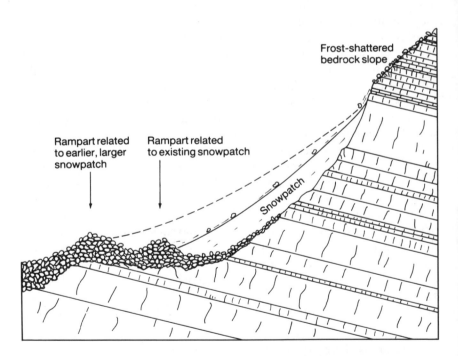

Figure 3.9 Diagrammatic representation of the formation of a protalus
rampart by rock fall over a snowpatch (after Catt, 1986a, figure 2.20).

quarter and vary in length between 60 and 550 m and in height between 1 and 10 m (Oxford, 1985). Welsh examples are known from the Tal-y-Llyn valley (Watson and Watson, 1977), the Black Mountains, and Snowdonia where on the east side of the Nant Ffrancon valley the A5 runs for around 400 m along the base of one of the finest examples (Gray, 1982). Ellis-Gruffydd (1977) maps the location of 13 protalus ramparts on the Old Red Sandstone escarpment of South Wales.

There is a general consensus that British protalus ramparts formed during the Loch Lomond advance, and this view is based on three arguments (Ballantyne and Kirkbride, 1986): (1) all recorded examples occur within the limits of the Late Devensian glacial maximum so that the ramparts developed after the decay of that ice sheet; (2) the Loch Lomond Stadial is the only established phase subsequent to the Late Devensian glacial maximum that was cold enough to provide the perennial snowpatches required for rampart formation; (3) virtually all documented examples lie outside (often immediately outside) the mapped limits of Loch Lomond advance glaciation.

All British ramparts possess only a single ridge or ramp crest, whereas in some other parts of the world multiple-ridge ramparts have been described. Ballantyne and Kirkbride (1986, p. 665) suggest that this indicates that the melting of the snowbeds responsible for rampart formation was uninterrupted by any significant return of colder or snowier conditions after the maximum of the Loch Lomond Stadial.

Avalanche chutes and dells

The northward-facing scarp of the Chalk along the Berkshire Downs reveals the existence of highly localized, short, narrow, parallel or subparallel, seldom-branching, steep gulleys. These occur either as crenulations on unbroken scarp faces or as serrated indentations on scarp-face dry valleys. The features have an average spacing of just under 50 m, are generally around 100 m long, and have a maximum slope angle of 30–36°. They only occur on the steepest parts of the scarps as, for example, on the slopes of the Manger near the Uffington White Horse (Goudie et al., 1980).

The origin of these 'washboard' slopes is far from certain, but they are morphologically very similar to the *coulées* of South Alberta

(Canada), which are formed by wind-driven snow, and to the avalanche chutes (*rasskers*) of Fennoscandia. It is possible, therefore, that they are further evidence of the power of periglacial nival processes in southern Britain.

Other minor features that may have a periglacial origin are dells. These are amphitheatrical shaped hollows, 100–200 m across; they are perched on valley sides, sometimes singly, in other places in groups tributary to small dry valleys, and are generally 5–9 m deep. They are well displayed on Eocene sands and clays in the New Forest of Hampshire (figure 3.10). The superficial material on their floors and at the foot of the steep slopes on which they are developed, is solifluction debris (Tuckfield, 1986). Other examples occur in the Keuper Marl of south-east Derbyshire (Jones, 1979).

Rather larger than dells, though possibly genetically related to them, are dry valleys. These may also have a periglacial origin and are discussed at length in chapter 5.

Cambering and valley bulging

Many valley-side and scarp slopes in Britain show a variety of superficial structures which have often been attributed to periglacial conditions. These structures which come in a variety of different forms are called cambers, ridge-and-trough features, gulls, valley bulges, etc.

Cambering is a process whereby the local dip of rock strata becomes much greater than the regional dip. Normally gently dipping strata have had their joints opened up to give deep cracks (gulls), and rafts of rock have moved downwards in the terrain opening up a series of troughs. Disturbed strata swathe the hill tops and choke or drape the hill slopes. Figure 3.11 provides a diagrammatic illustration.

The classic location for such structures is the Northamptonshire Ironstone Field (Hollingworth et al., 1944), but they are extensively developed in the Cotswolds (Ackermann and Cave, 1967) on the Corallian scarp near Oxford (Arkell, 1947); the Corallian of North Yorkshire (Cooper, 1980); in the East Midlands (Kellaway and Taylor, 1953); on the Hythe Beds and Lower Tunbridge Wells Sand in the Weald (Shepherd-Thorn, 1975; Bristow and Bazley, 1972; Worssam, 1963); near Barnsley, Yorkshire (Shotton and

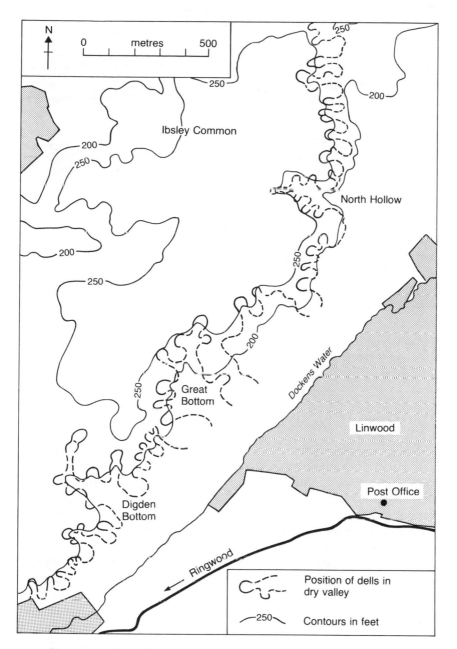

Figure 3.10 The distribution of dells on the east slope of Ibsley Common in the New Forest of Hampshire. Dells are amphitheatrically shaped valley side hollows (after Tuckfield, 1986, figure 3).

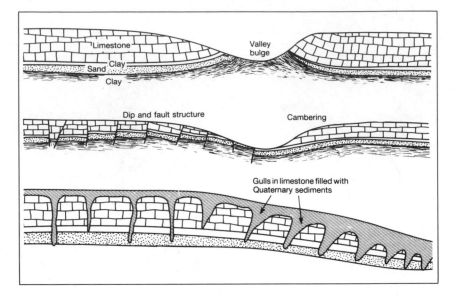

Figure 3.11 Structures, including valley bulges, cambering, gulls and dip and fault structures, associated with frost action in valleys cut in approximately flat-bedded sedimentary rocks of variable lithology (from Catt, 1986, figure 3.28).

Wilcockson, 1950); and in the Bunter Sandstones of Derbyshire (Jones, 1979).

Valley bulging, which often occurs in association with cambering, is particularly well developed in the British Jurassic and Cretaceous sedimentary rocks which contain thick argillaceous beds (Horswill and Horton, 1976). The rocks affected by valley bulging are intensely fissured and commonly sheared and have obviously been affected by intense pressures which have resulted in upward displacement of the rocks.

The clearest surface manifestations of such superficial structures are the ridge-and-trough features of the North Cotswolds (figure 3.12). Complexes of ridges and intervening troughs occur in the valleys of the Windrush and its tributaries, running parallel or subparallel to the main valleys, across or oblique to the line of greatest slope (Briggs and Courtney, 1972). They are up to 8 m deep and can be over 250 m long. Other clusters occur on the edge of the Cotswold escarpment at Cleeve Hill near Cheltenham (Goudie and Hart, 1975).

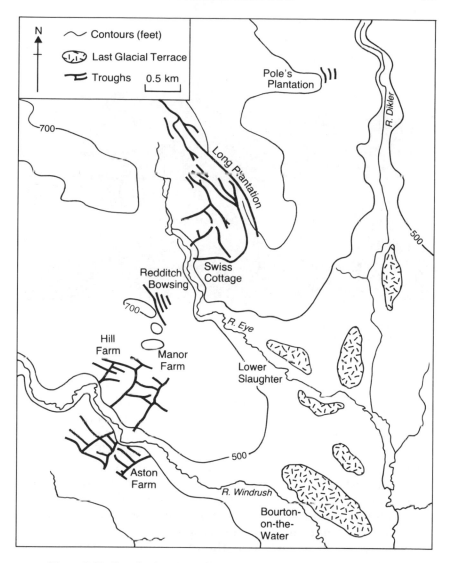

Figure 3.12 Gravitational slip trough features developed in the Jurassic beds of the Cotswolds near Bourton-on-the-Water (from Briggs and Courtney, 1972, figure 1).

Various hypotheses have been postulated for the origin of cambering and valley bulge structures. Hollingworth et al. (1944, p. 30) believed that 'the varied superficial movements . . . have a common factor in their origins in that their development is

Plate 7. One of the most impressive slip troughs on the Cotswold escarpment, occurs on the southern end of Cleeve Hill, overlooking Cheltenham. The whole hillside has slipped and rotated to produce a deep, linear trough. (Author)

primarily determined by the behaviour under the influence of erosion of relatively competent rocks when resting on the thick bed of clay.' Thus differential unloading resulting from valley downcutting was seen as a critical factor. Subsequently, Kellaway and Taylor (1953) suggested that an association with permafrost may have been essential. Certainly, downcutting of streams might be considerable under a nival regime in which run-off is encouraged by the presence of subsoil permafrost. The presence of permafrost might also allow the lower incompetent beds to support material above until, following climatic amelioration these incompetent beds became saturated and tended to deform. More recently, Kellaway (1972) suggested that some of the structures are primarily due to glacial erosion and subsequent unloading of the ice sheet. It is certainly possible that the loading and unloading caused by ice-sheet growth and decay could play a role (Horswill and Horton, 1976) and that subglacial drainage channels could carve through the competent into the incompetent beds. However, this mechanism would not apply to structures beyond the accepted limits of glaciation (e.g.

those in the Weald). Another possible contributing factor is ice-wedge growth, which could facilitate gull growth (Worssam, 1963).

Altiplanation (cryoplanation) terraces

A combination of extreme frost shattering and effective transport of coarse clastic particles over low-angle slopes by solifluction is thought to be capable of producing a series of rock-cut terraces on hill-slopes under tundra conditions. Such terraces are generally called altiplanation or cryoplanation terraces.

The first geomorphologist to recognize altiplanation terraces in Britain was the Frenchman, A. Guilcher (1950), who described them from indurated arenaceous sediments on Holdstone Down, Great Hangman and Trentishoe Down in North Devon. They also occur in South Devon, on Dartmoor (Waters, 1962) where they have the following characteristics:

Height of scarps	2–12 m
Dimensions parallel to regional slope	10–90 m
Dimensions transverse to regional slope	up to 800 m
Slope of treads	3–8°
Slope of scarps	15–22°
	(with occasional vertical rock outcrops)

On Dartmoor they are especially well displayed on the more well-jointed rocks of the metamorphic aureole (metamorphosed sediments and metadolerites) (Gerrard, 1988). Other examples have been described from the north Pennines (Tufnell, 1971); from the shale outcrops of the Upper Derwent Valley in the South Pennines (McArthur, 1981); from the north side of Black Down, an Old Red Sandstone area in the Mendips (Te Punga, 1956); from the gritstone slopes of the Esk basin in north-east Yorkshire (Gregory, 1966); and from the Torpantau area of the Brecon Beacons (Lewis, 1970).

Tors

Tors, 'conspicuous and often fantastic features' (Linton, 1955), are another widespread landform that may owe a great deal to periglacial events but which have also been the subject of acute controversy. Following Pullan (1959, p. 54) we can define a tor as 'an exposure of rock *in situ*, upstanding on all sides from the surrounding slopes and . . . formed by the differential weathering of a rock bed and the removal of the debris by mass movement.'

Although tors are generally associated with the granite terrain of Dartmoor, in reality they occur on a wide range of rock types (table 3.2), and have a broad distribution within England and Wales (figure 3.13).

Linton (1955) envisaged tors as being developed as a result of deep chemical weathering penetrating along joints in granite (figure 3.14a) and producing an irregular weathering front, so that masses of sound or nearly sound rock project upwards out of

Table 3.2 Lithologies associated with British tors

Area	Lithology	Source
Dartmoor (Devon)	Granite	Linton (1955) Eden and Green (1971)
Exmoor (Valley of Rocks)	Sandstone	Mottershead (1967)
Weald (Kent, Sussex)	Ardingly Sandstone (Lower Tunbridge Wells Sandstone)	Robinson and Williams (1976)
Charnwood Forest (Leicestershire)	Granite, Microdiorite and Hornstone	Ford (1967)
Pennines (Derbyshire)	Millstone Grit and Gritstone	Cunningham (1965)
Derbyshire	Dolomitized Carbon- iferous Limestone	Ford (1967)
Bridestones (NE Yorkshire)	Passage Beds of the Corallian (silicified grits and calcareous beds)	Palmer (1956)
Cheviot Hills (Northumberland)	Granite	Common (1954)
Stiperstones (Shropshire)	Quartzite (Ordovician)	Goudie and Piggott (1981)
Pembrokeshire (Wales)	Flinty Thyolite	Linton (1955)
Prescelly Hills (W Wales)	Dolerite	Linton (1955)
Scilly Isles	Granite	Scourse (1987)

Plate 8. Tors on non-granite lithologies: above, the gritstone tors of Ramshaw Rocks; below, the Stiperstones (Ordovician quartzite). (Author)

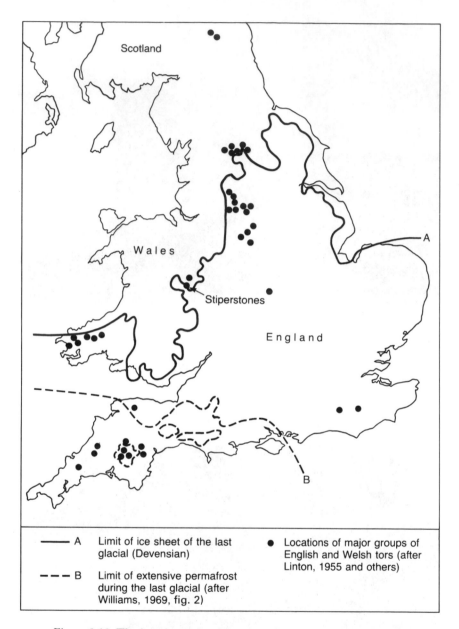

—— A   Limit of ice sheet of the last
        glacial (Devensian)

— — — B   Limit of extensive permafrost
        during the last glacial (after
        Williams, 1969, fig. 2)

●   Locations of major groups of
     English and Welsh tors (after
     Linton, 1955 and others)

Figure 3.13 The locations of major areas of tors in relation to the ice limit and permafrost distribution of the Devensian.

surrounding regolith. Such weathering, he contended, took place under the relatively warm, humid phases of the late Tertiary (see p. 19). The weathered regolith is then stripped away, perhaps by meltwater or solifluction during the Pleistocene, exposing the more massively jointed parts of the granite as tors and rounded corestones. It needs to be pointed out, however, that such a two-cycle deep weathering mechanism would not be applicable to rocks which are not subject to extreme chemical decomposition (e.g. the Ordovician quartzites of the Stiperstones). Moreover, recent work by Doornkamp (1974) using an SEM, and Eden and Green (1971) using mineralogical analysis of quartz/feldspar ratios, has indicated that relict patches of growan (which Linton has interpreted as being due to extreme chemical decomposition) do not have the appropriate characteristics to be expected if they had indeed been subjected to extreme chemical attack.

Palmer and Neilson (1962) have championed an alternative, 'periglacial' hypothesis (figure 3.14B) suggesting that Linton's deeply altered material is highly localized and the product of pneumatolysis. They believe that intense Pleistocene frost-shattering in jointed rock, and the effective down-slope removal of debris by solifluction, would cause more resistant, massively jointed compartments of rock to be left upstanding as tors, and for the debris to be evacuated to give angular boulder cover (clitter) to slopes. Certainly, the presence of large spreads of angular clitter at great distances from the tors, and often sorted into patterned ground, is suggestive of the power of this mechanism (Gerrard, 1988). For example, Goudie and Piggott (1981) have described boulder stripes and polygons around the Stiperstones of Shropshire, while similar features are known from parts of Dartmoor (e.g. Hen Tor, Harter Tor, Cowsend Beacon) (Palmer and Neilson, 1962).

It is possible, however, that some tors are still forming under present-day conditions even if Linton (1955, p. 470) dismisses such a notion, finding that 'present-day atmospheric weathering is acting upon them destructively and not constructively. Many tors are conspicuously in a state of collapse.' Palmer (1956) believed that the Bridestones of north-east Yorkshire were still being formed from the disintegration of a resistant stratum following the rejuvenation of a mature hill-slope. Certainly, jointing and the presence of zones of resistant rock are prerequisites for most tors and, over time, a whole series of weathering and erosional mechanisms may have operated to etch out lithological and structural differences to produce tors. It is for this reason that

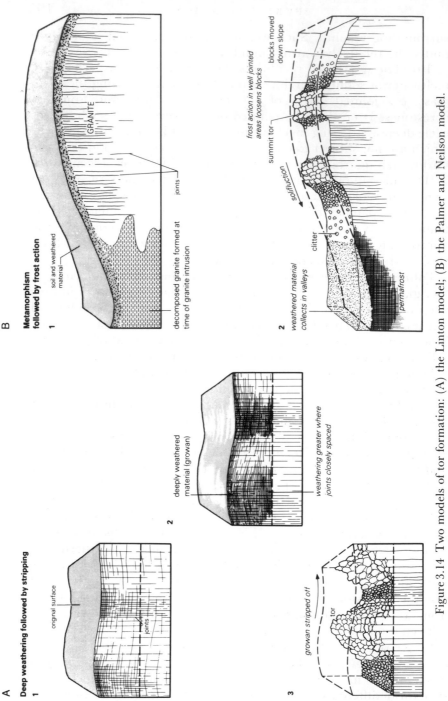

Figure 3.14 Two models of tor formation: (A) the Linton model; (B) the Palmer and Neilson model.

Gerrard (1974) produces what is essentially a compromise view of tor evolution (p. 48):

> The formation of many of the tors would then seem to have been a continuous process from the mid-Tertiary period onwards. The unloading domes would have formed during successive periods of planation when something approaching a type of etchplain may have existed. A reconstruction of the present distribution of weathered material and exposures of sheeted granite is very similar to incised etchplains described in other parts of the world. Tors would probably have developed as further incision opened up the vertical joints to the action of chemical weathering agents and would accord with the ideas of Linton. Subsequently during the cold periods of the Pleistocene the summit tors would have been modified by cryo-nival processes.

Loess and coversands

Although nowhere does it attain the thickness or extent that the loess deposits of neighbouring parts of Europe achieve (such as Normandy, the Paris Basin, and Germany), loess does occur in Britain (table 3.3). It is primarily patchy and thin in character,

Table 3.3 Selected reported examples of loess in England and Wales

Location	Source
Devon	Harrod et al. (1973)
Lizard, Cornwall	Roberts (1985)
North Norfolk	Catt et al. (1971)
East Yorkshire and Lincolnshire	Catt et al. (1974)
Lancashire (Morecambe Bay)	Boardman (1985), Vincent and Lee (1981)
Weald	Burrin (1981)
Wirral, Cheshire	Lee (1979)
Pegwell Bay, Kent	Wintle (1981)
Durham	Trenchmann (1920)
Cornwall	Catt and Staines (1982)
General	Catt (1979)

and, being mixed by frost churning into subjacent deposits, it is often more a contaminant of soils than a major, discrete band of pure sediment. Originally more extensive, it has been removed from many areas by such processes as solifluction, stream erosion and colluviation resulting from Holocene deforestation and agriculture (Catt, 1977). Loess is best preserved on limestone and dolomitic limestone, possibly because of enhanced cementation as a result of the close proximity of lime, and partly because it is less easily eroded by water action on permeable limestone surfaces in comparison with shales, clays, etc. On other lithologies much of the loess has been eroded to contribute to the silty, alluvial fills of many English river valleys (Burrin, 1981). The general distribution pattern is shown in figure 3.15. It is evident that the most imposing loess deposits in England occur over river terrace gravels, Tertiary sediments or the Chalk on either side of the Thames estuary. The classic location is Pegwell Bay, near Ramsgate.

Although deflation from glacial outwash and moraine under dry glacial and periglacial conditions doubtless provided conditions for loess supply on numerous occasions in the Pleistocene (as is evident in the thick loess sequences of Europe and Asia), very little old loess is preserved in Britain. So, while there are some isolated deposits of pre-Ipswichian loess (e.g. on the Chiltern Hills, and in the Thames and Gipping Valleys) most deposits are of Late Devensian age (Wintle, 1981; Parks and Rendell, 1988). There are, however, exceptions to this generalization, including the Barham Loess of northern Essex, which is overlain by Chalky Boulder Clay and is probably of Anglian age (Rose and Allen, 1977). Even the late Devensian materials are thin, weathered throughout, and mixed with other materials by slope processes and stream activity, but their ultimate aeolian origin has been indicated by mineralogical and grain-size affinities with deposits that do fit the normal, strict definition of loess. None the less, there is some confusion surrounding the definition of true loess and its differentiation from other silty deposits often loosely referred to as 'brickearth' (Catt, 1979).

The loess at Pegwell Bay attains a thickness of around 4 m, but this is an exceptional figure for Britain. British loess is generally less than 1 m thick.

Another type of aeolian deposit of Pleistocene age which has some significance is dune material, composed primarily of sand-sized sediments, which is generally termed 'coversand'. The distribution of coversands is shown in figure 3.15.

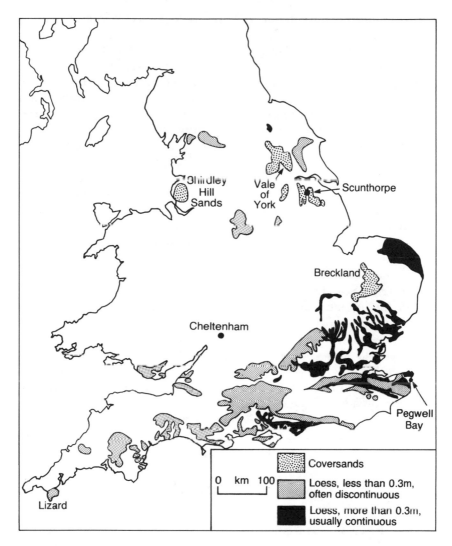

Figure 3.15 The distribution of loess deposits and of aeolian coversands in England and Wales (modified from Catt, 1977, figure 16.1).

Such aeolian sands occupy the southern parts of the Vale of York and the Lower Trent Valley, including a substantial area to the south of the Humber estuary in the vicinity of Scunthorpe. Other coversands occur in south-west Lancashire (the Shirdley Hill Sands), and in the Breckland of East Anglia. Along the Cotswold escarpment in Gloucestershire there are local accumulations called the Cheltenham Sands. Heavy mineral analyses indicate that they were probably derived by winnowing of the Main Terrace of the Severn some time during, or after, the Upton Warren Interstadial of the Mid-Devensian (Briggs, 1975). In Yorkshire and Lancashire the coversands occur above Late Devensian till, and radiocarbon dates from Lincolnshire suggest that the aeolian phase occurred after 11,000 BP forming well defined dune ridges (figure 3.16). The Shirdley Hill Sands, which blanket more than 200 km^2 of the south-west Lancashire coastal plain to a depth of 1–3 m are of Late Glacial age (Zone III of the late Devensian) and Wilson et al. (1981) have established the provenance of the sand from fluvioglacial deposits by wind and water sorting.

Further evidence for the importance of Pleistocene wind action comes from the presence of wind-sculpted clasts on certain Quaternary surfaces. Triccas (1973, p. 203) for example, found numerous ventifacts in the Midlands, 'associated with the red tills of the Irish Sea (Devensian) glaciation, either lying on the tills or on the fluvio-glacial terraces and often in close association with the patterned ground of periglacial origin. . .'. He postulated that in Late Devensian times katabatic winds moving off the decaying ice sheet would sweep over the bare tundra surfaces and outwash terraces blowing out the fine sands from between the gravels and leaving the ventifacts lying on the surface as part of the lag-gravels. Other spreads of Devensian ventifacts, from the north-east Cheshire Basin near Chelford, have been described by Thompson and Worsley (1967).

Present-day periglacial activity

Although permafrost is absent from contemporary Britain, it can be argued that upland Britain has an active periglacial environment (Ballantyne, 1987). Small-scale, active, sorted patterned ground has been reported from Snowdonia, the Lake District and the Pennines. Likewise active solifluction features and ploughing boulders exist (Tufnell, 1969, 1972; Ball and Goodier, 1970). Snow patches and nivation occur (Tufnell 1971; Vincent and Lee, 1982) and catastrophic avalanches happen from time to time. The most notable avalanche in southern England was that of Christmas Eve, 1836, when, following severe snow, an avalanche fell from the

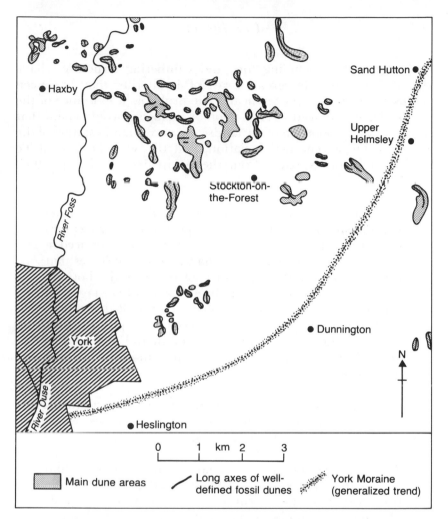

Figure 3.16 The distribution and trends of late Quaternary dunes near York (from Matthews, 1971, in Curtis et al. 1976, figure 7.2).

South Downs on to a group of cottages in Lewes, killing eight people. The scene of the calamity is marked by the present 'Snowdrop Inn' on the Eastbourne Road. Snow-lie can be prolonged in highland areas: 105 days in the year for Cross Fell (880 m) in the Pennines and 70 days in the year for Snowdon (900 m).

Conclusion

Although some, including tors and cambering structures, have disputed origins, the great array of landforms that has been discussed in this chapter indicates the probable importance of the tundra legacy. Indeed, certain geomorphologists would claim that a substantial proportion of the landscape was a manifestation of the power of Pleistocene periglaciation, and this was the tone of Te Punga's missionary paper from the Antipodes (Te Punga 1957, p. 410):

> The rapid wasting away of the land surface during periglaciation, due to the transportation of enormous quantities of material to lower levels, has produced a landscape of subdued aspect characterised by slopes that are convex near the top and concave near the bottom. The convex upper slope has been a zone of wastage and the concave lower slope has been a zone of deposition. Such slopes, developed during periglacial episodes, now dominate the present landscape of southern England. The general absence of features other than those of subdued form is most striking.

However, many of Te Punga's inferences are in fact still largely conjectural and a note against periglacial extremism has been expressed by Williams (1968, p. 311):

> The fact that many periglacial features can be found proves that the period had widespread effects: and the preservation of these features demonstrates that post-glacial erosion has been minor, at least where the features are found, though obviously it is not easy to know how many features have been removed. These considerations do not preclude the possibility that the slopes and drainage lines are mostly a legacy of Tertiary or interglacial conditions. The erosion of slopes involves the destruction of much if not all of the previous form. The last periods of erosion are necessarily better evidenced than the first even though their contribution to the development of the relief may have been small. Many of the phenomena that Te Punga lists, such as stone stripes

and involutions, are superficial ornamentations of the slope on which they occur and are not necessarily connected with the evolution of the relief.

We shall return to further possible consequences of periglaciation in chapter 5 when we discuss the formation of dry valleys, asymmetric valleys and river terrace aggradation.

4

Rocks and Relief

Introduction

Glaciation and periglaciation have in a sense done little more than embroider a pattern of relief that owes its gross form to the nature of underlying rocks. Their character and their disposition have a basic and profound effect on the landforms of England and Wales. So, for example, Mackinder (1904) divided Britain into two parts according to whether or not the rocks that outcropped at the surface were older or younger than the end of the Carboniferous (figure 4.1). A line drawn between the Tees and the Exe approximately delimited these two provinces. To the north and west of the line outcrops are the older, more resistant, and often highly deformed rocks of 'Highland Britain'; while to the south and east the rocks are for the most part sedimentary rocks which have been only gently tilted or relatively slightly folded. They underlie 'Lowland Britain' and have been etched out by denudational processes to yield a succession of 'scarp-and-vale' scenery (Jones, 1985), most of which lies beneath the 300 m contour.

The division made by Mackinder, as he himself recognized, is in reality more complex, for the boundary between the rocks of Highland and Lowland Britain is irregular rather than a straight line (figure 4.2) and in certain parts of England splinters of old rock emerge through the younger cover rocks (e.g. Charnwood Forest in Leicestershire, see p. 140).

None the less, the division provides an important starting point for understanding some of the most salient differences of the

Surface of Carboniferous and older rocks.

Surface of rocks newer than Carboniferous,
underlain by Carboniferous and older rocks.

Figure 4.1 The distribution of the older and new rocks of Britain as
represented by Mackinder (1904).

landscape. As Mackinder wrote (1930, p. 63) in his *Britain and the British Seas*:

> The contrast between the south-east and the north-west of Britain, between the plains and low coasts towards the continent, and the cliff-edged uplands of the oceanic

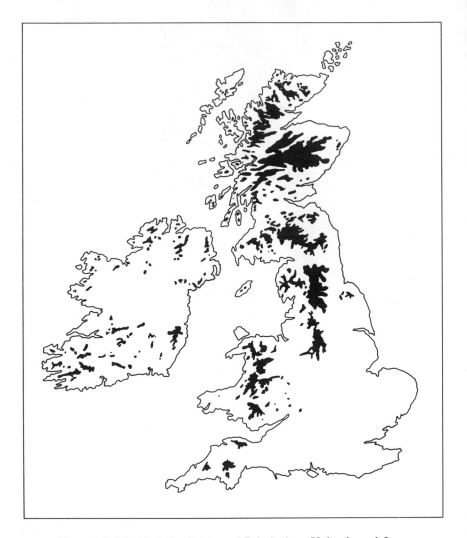

Figure 4.2 Mackinder's division of Britain into Uplands and Lowlands. The shaded areas are those of greater elevation than 1000 feet (*c*.300 m).

border . . . depends on a fundamental distinction in rock structure. In the south-east are sands, clays, friable sandstones, chalk and soft limestones. In the north and west the sandstones are for the most part hard and gritty, often so hard and compact as to become quartzite; the clays have been pressed into slates; the limestones are crystalline, sometimes becoming marble, and igneous rocks abound, which, together with the sediments, have often been crushed into schist and gneiss. In Britain . . . the harder rocks are the more ancient, and despite the lapse of time, they still rise into uplands and into peaks, while the newer softer deposits of the south-east have been reduced to lowlands in a comparatively short geological time.

On the basis of rock types and rock arrangement it becomes possible to understand the great variety of relief in England and Wales and to appreciate the geomorphological regionalization shown in figure 4.3. As A. C. Ramsay said in 1878 (p. 302), 'England is the very Paradise of Geologists, for it may be said to be in itself an epitome of the geology of almost the whole of Europe, and much of Asia and America'.

The physical properties of the rocks

The qualitative differences which Mackinder identified between the geomorphological performance of different types of rock can now be taken further, because some basic rock mechanics data are now available for some of the major rock types. These data are listed in table 4.1.

By ranking these data so that those rocks with the greatest density, lowest porosity, greatest unconfined compressive strength, greatest hardness and greatest modulus of elasticity are ranked as the most 'resistant' rocks, one can get some feel for the likely influence the various types of sedimentary rock will have on relief (table 4.1A). From this table it is apparent that there is a very wide range of values for sedimentary rocks, with the Upper Chalk and the Bunter Sandstone being the 'weakest' and the much older Bronllywyn Grits and Horton Flags being the 'strongest'. The Carboniferous Limestone emerges as the strongest of the limestones.

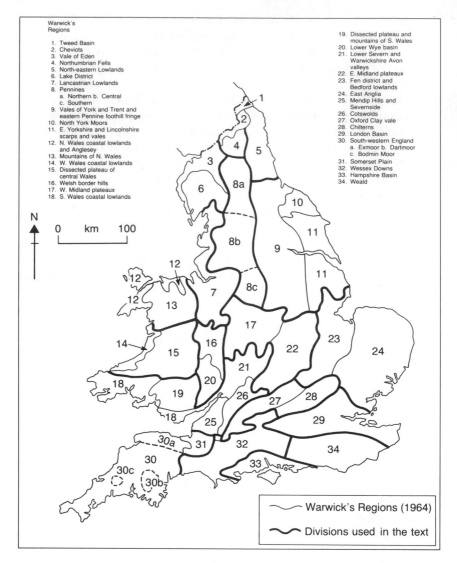

Figure 4.3 The relief regions of England and Wales, showing Warwick's (1964) delimitation.

However, in most respects these sedimentary rocks have considerably lower strengths than the ancient metamorphic and igneous rocks (table 4.1B). The compressive strengths of the latter are on average almost three times those of the sedimentary rocks, the surface hardness about 1.8 times, and the Young's modulus about 2.2. times.

More data of this type are required before firmer conclusions can be drawn about the precise relationships between measured rock properties and relief. None the less the general relationships between rocks and the relief regions depicted in figure 4.3 can now be discussed.

Cheviot Hills, Northumbrian Fells and north-eastern lowlands

The Cheviot hills of northern England, which lie along the Scottish border, are composed of a pile of andesitic lavas which were extruded in Devonian times (345–395 million years ago) on to a Silurian platform. Granite was subsequently intruded and a blanket of Carboniferous sediments (280–345 million years old), since eroded away, applied as a covering. The granite core now forms the twin summits of the Cheviots (815 and 716 m OD). The Cheviot dome is characterized by rounded, peat-covered slopes, and was glaciated in the Pleistocene, when many meltwater channels were formed (see p. 59).

Around the Cheviots is a belt of upland terrain formed in the Carboniferous rocks that have been stripped from the Cheviot Dome. The Carboniferous rocks, dipping generally southwards and eastwards, descend in the south to the great structural trough of the Tyne Gap. The Carboniferous strata produce complex cuesta and strike valleys, and this is a reflection of their variable lithology. The Cemenstone group of shales and muddy limestones and the Scremerston Coal Series of softer sediments tend to have been exploited by river erosion, while the massive, coarse Fell Sandstones form a crescent-shaped upland girdle around the Cheviots.

On the eastern side of the Pennines, Northumbrian Fells and Cheviots, there is a broad, predominantly lowland area. The Carboniferous Limestone and Millstone Grit formations of the Pennines dip eastwards and are succeeded as one moves towards the North Sea by the Lower Coal Measures, the Middle and Upper Coal Measures, the Magnesian Limestone and higher Permian beds, the Bunter Sandstone and the Keuper Marls.

Table 4.1 Basic rock mechanics data

Rock type	Dry density (mg/m³)		Porosity (%)		Dry unconfined compressive strength (MPa)		Schmidt hammer Hardness		Young's modulus (×10³ MPa)		Total score	Rank
A Sedimentaries												
Carboniferous Limestone	2.58	(3)	2.9	(2)	106.2	(3)	51	(3)	66.9	1	12	3
Bath Stone (Jurassic)	2.30	(4)	15.6	(7)	15.6	(8)	15	(8)	16.1	8	36	8
Middle Chalk (Cretaceous)	2.16	(7)	19.8	(8)	27.2	(7)	20	(7)	30.0	5	34	7
Upper Chalk (Cretaceous)	1.49	(10)	41.7	(10)	5.5	(10)	9	(10)	4.4	10	50	10
Fell Sandstone (Lower Carb.)	2.25	(6)	9.8	(4)	74.1	(4)	37	(4)	32.7	4	22	4
Chatsworth Grit (Carboniferous)	2.11	(8)	14.6	(6)	39.2	(6)	28	(5)	25.8	6	31	6
Bunter Sandstone (Triassic)	1.87	(9)	25.7	(9)	11.6	(9)	10	(9)	6.4	9	45	9
Keuper Waterstones (Triassic)	2.26	(5)	10.1	(5)	42.0	(5)	21	(6)	21.3	7	28	5
Horton Flags (Carboniferous)	2.62	(2)	2.9	(1)	194.8	(2)	62	(1)	67.4	2	9	2
Bronllywyn Grit (Cambrian)	2.63	(1)	1.8	(1)	197.5	(1)	54	(2)	51.1	3	8	1
Mean					71.37		30.7		30.21			

B *Metamorphic and igneous*

Mount Sorrel Granite	176.4	54	60.6
Eskdale Granite	198.3	50	56.6
Granophyre	204.7	52	84.3
Andesite	204.3	67	77.0
Basalt	321.0	61	93.6
Slate	96.4	42	31.2
Gneiss	162.0	49	46.0
Hornfels	303.1	61	109.3
Mean	208.23	54.5	69.82

Data processed from Bell (1983).
Ranks in parentheses.

The coastal plain of Cumbria, the Vale of Eden and the Lake District

The coastal plain of Cumbria lies between the Lake District and the Solway Firth. It is primarily underlain by Carboniferous Beds, with Carboniferous Limestone in the east of the area, and with extensive areas of the Whitehaven Sandstone Series. In addition there are outcrops of New Red Sandstone. The Carboniferous Limestone has for the most part exerted a feeble influence upon relief, though there are examples of crags and scarp and dip scenery near Clints, Lamplugh and in the Pardshaw–Eaglesfield district. The Whitehaven Sandstone forms relatively high moorlands and cliffs and the New Red Sandstone is responsible for the grand coastal scenery of St Bees Head. Glacial deposits have masked the details of rock sculpture to a considerable degree (Eastwood et al., 1931).

The Vale of Eden is a long established lowlying area for it contains Permian fan deposits, called 'Brockram', derived from the erosion of Carboniferous limestone. This characteristic was reinforced during the movement along the outer Pennine fault during the Tertiary, when the Permo-Triassic rocks were let down against the Carboniferous on the western downthrow side of the fault (King, 1976, pp. 25–6).

The Lake District (figure 4.4) is largely composed of a complex elongated dome of Ordovician and Silurian rocks and it is rimmed by outcrops of Carboniferous and Permo-Triassic age. Caledonian earth movements impose an approximate south-west to north-east trend, and three broad outcrops cross the area (figure 4.5). In the north, the highly contorted and sheared Skiddaw Slates of the Lower Ordovician give the relatively smooth mountains of Skiddaw (931 m) and Blencathra (868 m). The scenery tends to be less rugged than the second belt, the central Lake District, which consists of the Borrowdale Volcanic Series. These are lavas, agglomerates and tuffs of the Middle Ordovician and form England's highest point – Scafell Pikes (978 m). Striking glacial landforms have been produced. The third belt, in the south, extends from the Duddon estuary past Windermere. It is relatively low, with summits in the range 300–580 m, and is built of Silurian sedimentary rocks. These include some limestones, but are mainly shales, flags, sandstones and mudstones. Characteristically, they have smoothed slopes and underlie the gently dipping monoclinal

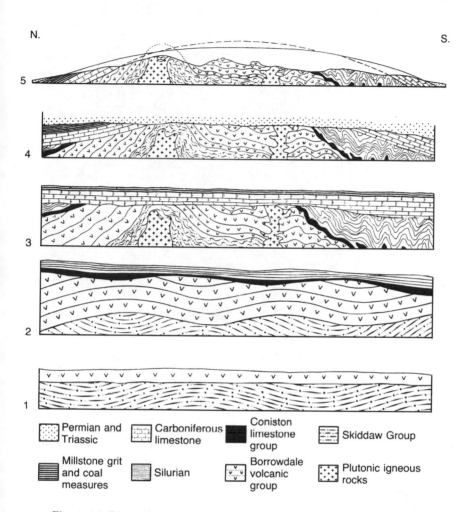

Figure 4.4 Diagrammatic sections to illustrate the building of the Lake District: 1 = deposition of Skiddaw Group; folding and erosion; deposition of Borrowdale Volcanic Group; 2 = folding and erosion; deposition of Coniston Limestone Group and Silurian rocks; 3 = severe folding and great erosion; intrusion of plutonic igneous rocks; deposition of Carboniferous rocks; 4 = gentle folding and considerable erosion; deposition of Permian and Triassic rocks; 5 = gentle uplift, producing an elongated dome and resulting in radial drainage; erosion to present form (from Taylor et al., 1971, figure 1).

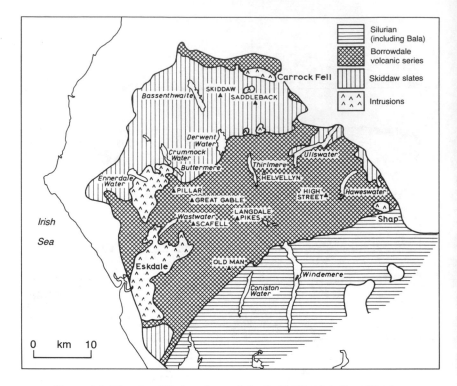

Figure 4.5 The simplified geology of the Lake District. Post-Silurian rocks remain unshaded (after Sparks, 1971, figure 7.6).

block of the Howgill Fells. Around this dome are younger sedimentary rocks which rest unconformably upon the Lower Palaeozoic core. Carboniferous limestones and dolomites produce striking escarpments to the south and east.

Within the Lake District there are some igneous intrusions, such as the Emmerdale granophyre and Eskdale granite to the west, and the Shap granite to the east.

A good early treatment of the geomorphology of the area is provided by Marr (1916).

The Lancastrian lowland

The Lancastrian lowland is an essentially lowlying area, where little land exceeds 135 m OD. Most of it is underlain by Triassic rocks (195–230 million years old) with a considerable cover of drift and, since Carboniferous times, it has been an area of intermittent subsidence and faulting. Thus the western margin of the Rossendales is sharply defined by a fault-line scarp, while downfaulted masses of Bunter Sandstone (Lower Triassic) form the highest ground in the Wirral and Liverpool. The outline of the Dee estuary also appears to be fault controlled, though glacial excavation has also played a role.

The Cheshire Plain is underlain by a broad synclinal structure; Keuper Sandstones (Upper Triassic) form scarps on its eastern edge, notably at Alderley Edge.

It is, however, probably naïve to see the relief simply in structural terms for at least three denudational cycles have been identified that are developed across a variety of both Permo-Triassic and Carboniferous Rocks (Freeman et al., 1966).

One notable feature of some of the lowland areas is the development of peaty mosslands. Comprehensive details are provided by Shimwell (1985).

The Pennines

Extending from the west Midland Plateaux to the Northumbrian Fells, the Pennines form the backbone of Britain, and constitute the main drainage divide between the North Sea and the Irish Sea. (figure 4.6). In general terms the Pennines form a monoclinal structure which tilts gently towards the east but is bounded on the west by abrupt faulting or downfolding. They are built mainly of Carboniferous rocks and comprise a series of dissected uplands which can attain heights in excess of 600 m.

In the south the Derbyshire Pennines, or Peak District, reach 636 m on Kinderscout. Erosion of an asymmetric dome has exposed the Carboniferous Limestone over a wide central area, 'The White Peak', and this is characterized by deep dales and extensive karstification. Particularly in the north-west, inward-facing escarpments of the Namurian Millstone Grit Series overlook the limestones. Massive beds of resistant grits give rise to high, peat-covered moorlands and sharp 'edges', but they are interbedded with and overlie shales (the Edale Shales) which frequently cause slope instability.

The Central Pennines are narrower and less elevated, and

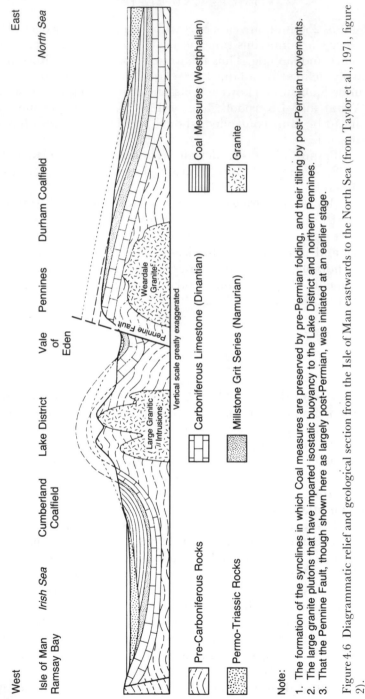

West

Isle of Man
Ramsay Bay

Irish Sea

Cumberland
Coalfield

Lake District

Vale
of
Eden

Pennines

Durham Coalfield

North Sea

East

Pennine Fault

Weardale
Granite

Large Granitic
Intrusions

Vertical scale greatly exaggerated

Pre-Carboniferous Rocks

Permo-Triassic Rocks

Carboniferous Limestone (Dinantian)

Millstone Grit Series (Namurian)

Coal Measures (Westphalian)

Granite

Note:

1. The formation of the synclines in which Coal measures are preserved by pre-Permian folding, and their tilting by post-Permian movements.
2. The large granite plutons that have imparted isostatic buoyancy to the Lake District and northern Pennines.
3. That the Pennine Fault, though shown here as largely post-Permian, was initiated at an earlier stage.

Figure 4.6 Diagrammatic relief and geological section from the Isle of Man eastwards to the North Sea (from Taylor et al., 1971, figure 2).

Plate 9. Brimham Rocks, Yorkshire – a modern view. (Michael J. Stead)

consist of the shales and sandstones of the Millstone Grit Series, flanked by Coal Measures to both east and west. They produce a somewhat tiered topography, with lower elevations in Coal Measure country. To the west of the Central Pennines the Millstone Grit gives rise to two smaller uplands which flank the valley of the Ribble: the Forest of Bowland (560 m) and the Forest of Rossendale (460 m).

Lying further north is the Askrigg Block. It is bounded to the south by the Aire Gap and to the north by the Stainmore depression. On the western and south-western edges the block is demarcated by major faults, including the Dent, Barbon and Craven faults. The Great Scar limestone (basal Carboniferous) gives some of the best karstic scenery in Britain and is particularly well exposed on the southern side of the block. More extensive in outcrop is the overlying Yoredale Series of limestones, shales and sandstones. This gives stepped topography and passes upwards into the Millstone Grit, which caps the summits of Whernside and Ingleborough. The upland areas are dissected by the Yorkshire Dales, river valleys of very variable character (King, 1960, p. 3): 'Some, like Swaledale, are steep and narrow, some are wider and more mature, such as Wensleydale. Others, like Bishopdale and Upper Wharfedale, are flat-floored and steep-sided, typically U-shaped, while Walden and Coverdale are V-shaped and have no wide flat floor.'

The northernmost upland block of the Pennine group is called the Alston Block. It is at Cross Fell (893 m) that the Pennines attain their highest elevation. Cross Fell Edge is a fault-line escarpment which marks the western boundary of the block, while to the south it is bounded by the Stainmoor Depression, and to the north by the Tyne Gap. Most of the block is composed of Carboniferous sediments, although in comparison with further south in the Pennines, limestones are less significant. Another important difference is the presence of a late Carboniferous dolerite intrusion, the Whin Sill, which locally forms cliffs, escarpments and waterfalls. Like the Askrigg block, the Alston block sinks gently eastwards beneath younger strata.

Eastern England from Tees to Wash

The western part of the region consists of the lowlying Trent Valley and Vale of York, both of which drain into the Humber. The Trent

Plate 10. Wharfedale is one of the Yorkshire Dales. Its broad form owes much to the influence of glaciation. (Tony Waltham)

Valley, in places 15 km wide, and mantled in Quaternary deposits, is mainly underlain by Mercia Mudstone (Keuper Marl) with a border of Liassic Clays to the east (Kent, 1980). The Vale of York is even wider, but so heavily mantled in drift that the solid rock exerts little direct influence on relief. None the less it is fundamentally a strike valley developed in unresistant Triassic sediments (Peel and Palmer, 1955). It narrows gradually northwards until in the vicinity of Northallerton the Jurassic escarpment approaches the Pennines.

The Trent Valley is overlooked to the east by the Lincoln Edge, an escarpment which is capped by Middle Jurassic Lincolnshire Limestone. Further east still is a clay lowland, the Lincolnshire Clay Vale, which extends from the Humber in the north to the Fenlands in the south. It is underlain primarily by Upper Jurassic Clays, with a carpet of drift. This vale in turn gives way eastwards to the Lincolnshire Wolds, a north-west-trending belt of dissected Chalk uplands. They continue north of the Humber as the Yorkshire Wolds, which end at the North Sea in the bold cliffs of Flamborough Head.

To the north and east of the Vale of York three other zones can be recognized. The Howardian Hills, which rise to 172 m, are an undulating area cut by a series of mainly east–west faults. The Vale of Pickering is a broad plain that is developed on glacial drift and Upper Jurassic mudstones, lying between the dip slope of the Corallian rocks to the north and the Chalk escarpment to the south. Thirdly, the Cleveland Hills and North York Moors, which rise to over 450 m, are a dissected plateau area formed mainly of Middle Jurassic sandstones and shales with Liassic clays forming the lower slopes.

North Wales

The mountains of North Wales – the Carnedds, the Glyders, the Arenigs, the Arans and Cader Idris – are arguably the finest terrain south of the Scottish border. This character owes much to the geological structure of the area, as explained by Fearnsides (1910, pp. 787–8):

> It is a district formed pre-eminently of the products of Ordovician volcanoes, and includes also the finest development of Cambrian sediments known in Europe . . . Topographically, it shows an adaptation of surface detail to underground geology which it is difficult to surpass. Structurally this district includes two great anticlinal folds pitching to the north-east . . . Scenically, however, it is not the anticlines which count, for it is along the intervening syncline that the hardened masses of eruptive rock stand out among the sediments, and, dissected out by denudation, form some of the most beautiful mountains of our land. The valleys along the softer Upper Cambrian sediments, and the hills determined by the massive lower Cambrian grits of the anticlinal cores, are also picturesque, but lack the fretted grandeur of the more varied mountains of Ordovician rock.

The Cambrian (510–570 million years ago) Harlech Grits, which consist of alternating shales and grits, are geomorphologically highly important; this applies especially to the Lower Cambrian Rinog Grits which, in the centre of the Harlech Dome, 'give rise to some of the most spectacular relief in Wales outside the areas of volcanic rocks . . . All in all the Harlech series add diversity to the general plateau landscape of Wales, which is often classed as monotonous' (Sparks, 1971, p. 272).

The west Wales coastal lowlands and the dissected plateau of central Wales

To the south of the Dovey estuary the rocks are largely Silurian mudstones with cores of interbedded grits. Structurally, the area centres on the Teifi anticline, which trends roughly parallel to the coast of Cardigan Bay, and which strikes through Plynlimon and the River Teifi. To the north-west is a breached coastal syncline and to the south-east is the Central Wales Syncline, both of which stand out as extensive areas of Aberystwyth Grits.

Much of Central Wales forms tracts of dissected plateau, which reach their highest point in Plynlimon (figure 4.7). In general terms it can be described as being underlain by the soft, contorted sediments of the Upper Ordovician and Silurian beds of the Central Wales syncline. Over extensive areas these rocks, for the most part shale, mudstones, flags, grits and conglomerates, have been planed off by Tertiary erosional events (see p. 6).

The Welsh Borderland and Lower Wye Basin

The Welsh Borderland is an area of complex geology and complex geomorphology. In the northern part of the area Breidden (366 m) is a remarkable laccolite of dolerite; Moel-y-Golfa (403 m) is an intrusion of andesite; the Long Mountain (408 m) is composed of Upper Silurian rocks; Corndon (513 m) is another dolerite laccolite; the Stiperstones are a ridge of Ordovician Arenig quartzites with distinctive tors; and the Long Mynd (517 m) is a plateau of highly dipping pre-Cambrian rocks, which include grits, flags and conglomerates. The Long Mynd itself is flanked by hills composed of outcrops of Pre-Cambrian volcanics (e.g. Pontesford Hill). Ordovician and Silurian rocks, including the Wenlock and Aymestry limestones, form a succession of ridges and valleys (figure 4.8).

To the south of Ludlow, around Leominster and Hereford, the landscape is more subdued and rolling, and is underlain by Old

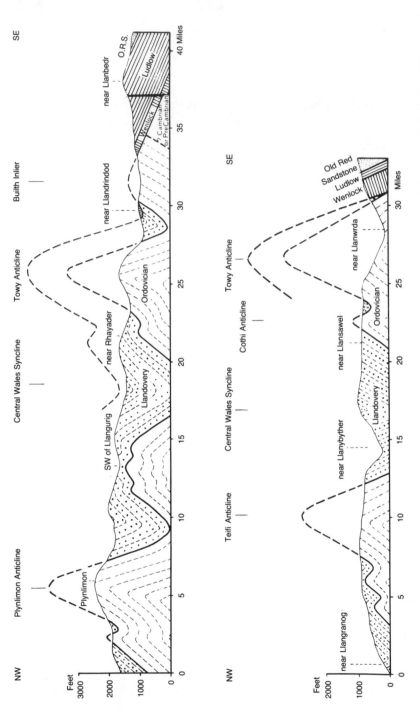

Figure 4.7 Generalized relief and geological sections across mid-Wales to show the development of the major Caledonian folds, and the intensity of pre-Wenlock (Silurian) earth movement (from George, 1970, figure 32).

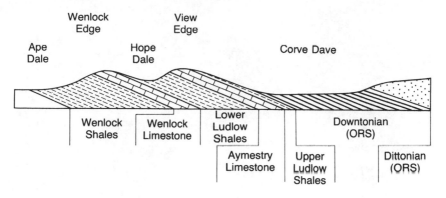

Figure 4.8 Diagrammatic section of relief and geology on the Welsh borderland (from Sparks, 1971, figure 7.8).

Red Sandstone (Devonian 345–390 million years old). However, diversity is provided on the margins by other rock types, of which perhaps the most notable are the Silurian, Cambrian and Pre-Cambrian rocks (diorites, schists, granite, etc.) of the Malverns (398 m).

The Malverns show the influence of structure with considerable clarity. The extremely steep slopes on either side of the ridge are associated with two major fault lines: the Malvern or Eastern Boundary Fault and the Western Boundary Fault. Transverse fractures are also well developed, and the fact that both the Herefordshire Beacon mass and that of Chase End Hill lie out of alignment with the rest of the range may be attributed to thrust faulting, while all the major cols and the Gullet through valley relate to cross-faulting (Morris, 1974).

Between the Severn Valley and the plateaux and mountains of South Wales there is a small but highly complex area that flanks the lower course of the River Wye. The river itself cuts deeply to form some of the finest incised and abandoned meanders in the British Isles.

Within the area is the Forest of Dean. Here Carboniferous rocks occupy a basin within the Old Red Sandstone. The coalfield is rimmed by the Carboniferous Limestone, and Pennant Sandstone crops out extensively between Coleford and Cinderford.

To the north and north-west of the Forest of Dean folding brings Silurian rocks to the surface creating the famous Woolhope Dome. Here the relationships between structure, lithology, landform and drainage are clearly displayed (figure 4.9).

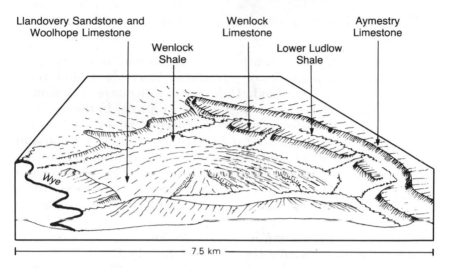

Figure 4.9 The Woolhope Dome developed in folded Silurian rocks of the Welsh borderland (from Dury, 1986, figure 25).

A resistant core of Llandovery Sandstone and Woolhope Limestone is encircled by weak Wenlock Shale, strong Wenlock Limestone, weak Lower Ludlow Shale, and strong Aymestry Limestone. Annular valleys have been incised into the shales, while the limestones create rings of higher ground. As Murchison (1854, p. 118) expressed it:

> . . . the denudation of the valleys which lie between the ridges that encircle the central dome has been so complete as to render it the finest known example, within the British Isles, of a valley of clean denudation as well as of elevation. Not only have no extraneous loose materials been translated to it from other tracts, but every fragment derived from the mass of rocks which must once have arched over it, has been swept out of the central and encircling hollows; a striking proof of the forcible agency exerted in the denuding operation.

Silurian sandstone also gives rise to the prominent May Hill.

Plate 11. On the Welsh borderland, the River Wye is deeply entrenched at Symonds Yat. (Tony Waltham)

The South Wales coastal lowlands, plateaux and mountains

The main features of the South Wales coastal lowlands is that they are broken up into a series of peninsulas. In Pembrokeshire there are the Pencaer, Dewisland, Dale and Castlemartin peninsulas, while in Glamorgan there is the Gower peninsula. In the northern part of the area Caledonian trends (north–east to south–west) dominate and the rocks are predominantly of Pre-Cambrian and Lower Palaeozoic age. In the southern and eastern part of the region the rocks are of Upper Palaeozoic and Mesozoic age and were folded in Hercynian times. They are the same rocks which make up the northern rim of the westernmost part of the South Wales coal basin.

The Vale of Glamorgan is a generally lowlying area stretching between the River Kenfig (in the west) and the Rhymney (in the east). Its northern boundary is formed by the south-facing scarp of hard Pennant Sandstone. In a line stretching between Cowbridge and Cardiff there is a central anticlinal feature which brings

Plate 12. The coastline at Castlemartin, South Pembrokeshire. The structures on this stretch of coastline are highly complex and produce a series of stacks and arches. Above the cliffs is a broad platform of erosional origin which truncates the strata. (Tony Waltham)

Palaeozoic rocks (including Carboniferous Limestone and Old Red Sandstone) to the surface. To the north of this structure there are Triassic and Liassic beds, while to the south there are alternations of the Lias limestones and shales. Much of South Wales is formed of a broadly oval-shaped plateau defined by bounding escarpments of the Carboniferous and earlier beds (figure 4.10).

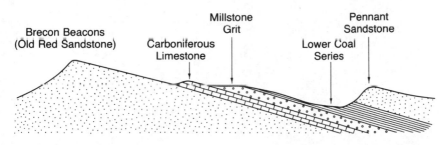

Figure 4.10 Diagrammatic section of relief and geology through the South Wales coalfield (from Sparks, 1971, figure 9.7).

The Pennant Sandstone Series occupies the greater part of the plateau surface and along the north and north west it gives rise to a prominent bounding escarpment formed of various units (e.g. Mynydd Sylen, Mynydd Betturs and Craig-y-Llyn).

Beyond this escarpment the Lower Coal Series strata occupy a broad depression, occupied by the broad, open valleys of the Gwendraeth Fawr and Amman. Further east, the heads of such valleys as the Cynon, Taff and Ebbw Fawr have a broader and more open appearance than where the valleys are developed in the Pennant grits.

Further north, beyond the margin of the South Wales coalfield is another series of north-facing escarpments culminating in the Old Red Sandstone masses of the Black Mountains, Fforest Fawr and the Brecon Beacons.

The South Wales Coalfield syncline is asymmetric: along the northern rim the rocks dip gently towards the south, while along the southern rim they dip very steeply to the north.

Plate 13. The Black Mountains escarpment, developed in the Old Red Sandstone of South Wales. (Cambridge University Collection: copyright reserved)

The Severn and Avon valleys, the Mendips and the Cotswolds

Sandwiched between the Welsh borderlands on the west and the Cotswolds in the east is one of the greatest clay vales in England and Wales. The western portion is underlain by Keuper Marl (Triassic) and the eastern portion by Lower Lias strata. Rhaetic beds of greater than average resistance create small escarpments that give some variety to otherwise rather featureless plains. Pleistocene terrace deposits are also widespread.

The Bristol-Mendip region is a country of varied relief, interesting drainage (Frey, 1975) and varied lithology, lying between the estuary of the Severn and the Jurassic scarplands. It has certain of the characteristics of the west Midland plateaux (see p. 140) in that it consists of 'islands' of older rocks wrapped round by the later and softer Triassic and Liassic deposits. The 'islands' are remnants of Hercynian Folds and are composed of rocks of varied age: the large Carboniferous Limestone masses of the Mendips, with its core of Old Red Sandstone, Coal Measures and some patches of Silurian material (e.g. in the Tortworth and Eastern Mendip Inliers).

The Cotswolds, which at their highest point near Cheltenham exceed 300 m in height, are a classic example of cuesta topography. They rise above the Severn Plain as a prominent escarpment. The Plain itself is underlain by incompetent blue Lias clays. Within the Middle Lias there are occasional harder beds of Marlstone which forms ledges in the main escarpment and which sometimes cap small outliers (e.g. Churchdown, Dixton, Battledown, Robinswood and Dumbleton Hills).

The main escarpment of the Cotswolds, which faces north-westwards, is much embayed, displays signs of landslide and debris flow activity, has many cambering structures, and also has a series of prominent outliers capped by Inferior Oolite (of which Bredon Hill is the largest example, and others are Woolstone, Oxenton, Langley and Ebrington hills). The cuesta is composed of the Inferior and Great Oolites of the Middle Jurassic, which have a regional dip of only 1–2°, and the dip slope is drained by various tributaries of the Thames, many of which provide fine examples of valley meandering. Some rivers display semi-karstic characteristics, losing discharge as they traverse the limestones, and some may go underground for some of their courses (e.g. the Leach) but, in general, the Oolitic limestones are not noted for their karstic development in the Cotswolds. North-eastwards into the East Midlands, as the lithology of the Jurassic beds changes, the cuesta becomes less well developed (see p. 140).

The Midland Plateaux

In the heart of England, between the Pennines, the Welsh massif and the Jurassic scarplands, is the great area of the west Midlands Plateaux. The most important of the geological formations in the area is the Upper Trias, consisting of Keuper Marls with the fine-grained Keuper Sandstones. There are also Lower Trias Bunter Sandstones and pebble beds and these give some high country, such as the Cannock Chase Plateau. A detailed discussion of the area, and of its horst-like structures and ancient inliers is provided by Clayton (1979). The Triassic beds (192–230 million years old) wrap around masses of older rock, which give a series of 'islands', including Charnwood Forest (with pre-Cambrian rocks), the Nuneaton Ridge (Cambrian Hartshill Quartzite), the Lickey Hills (also Cambrian Quartzite) and the various coalfields (Leicestershire, Warwickshire and South Staffordshire). Of these islands Charnwood is the most distinctive. As Mackintosh (1869, p. 349) wrote: '. . . it forms a very unexpected break in the general character of the scenery of the midland counties. It is wildness located in the midst of tame luxuriance . . . It looks like a part of Wales cut off and planted in the midst of "merry England" – an outlier of the great mountain territory of the principality.'

The East Midland Plateaux form a moderately elevated area that runs north-eastwards from the Cherwell Valley (between Banbury and Oxford) towards the erstwhile county of Rutland. They are formed entirely of Jurassic strata (140–195 million years old) and are bounded on the west by the Lias vales and on the east by the Oxford Clay lowlands. The relief is relatively low in comparison with other portions of the great Jurassic escarpments and the escarpments are ill developed. This is because thick beds resistant to erosion and weathering are poorly developed. So, for example, in Northamptonshire, the Inferior Oolite is represented by the Northampton Sand (ironstones and calcareous sandstones generally around 8 m thick) and by the even thinner Lower Estuarine Series (sands, silts and clays). In the Cotswolds near Cheltenham, the equivalent beds are predominantly limestones, over 100 m thick. However, the Middle Lias Marlstone, west of Grantham, produces a relatively bold escarpment overlooking the Vale of Belvoir, and is the main relief former in this region.

Also important in explaining this subdued relief is the decline in the regional dip, which in Northamptonshire is often as low as 5–6 m km^{-1}. As Clayton (1979, p. 153) explains:

> This low angle means that the high cuesta of the Cotswolds is replaced by a complex plateau area where at first sight little more can be established other than a general correlation between clay-floored valleys and more resistant (sandstone- or limestone-capped) ridges. A small-scale map will show a belted arrangement of the outcrops, but these are very wide and greatly complicated by the area of older beds exposed in all the main valleys.

Another feature of this area is the great extent of cambering (see p. 96) and landslipping (see p. 200).

The Oxford Clay lowland and the Chilterns

The Oxford Clay lowland is just that – a predominantly lowlying zone between the Cotswolds and the chalklands of the Berkshire Downs and the Chilterns, which is floored for the most part by Upper Jurassic Oxford Clay. The Clay itself is locally and quite extensively obscured by Pleistocene terrace gravels of the Thames and its tributaries, especially between Lechlade and Oxford. Elsewhere the Clay is exposed at the surface, creating large areas of ill-drained terrain of which Otmoor (just to the north-east of Oxford) is the most notable example. Downdip, towards the Chilterns, other clays promote lowland development, including the Kimmeridge and Gault clays, but they are separated from the Oxford Clay Vale proper by a discontinuous and rather unprepossessing little scarp formed in the Corallian strata.

To the north-east of the Goring Gap, where the Thames punches a way through the Chalk escarpment, the chalk cuesta is called the Chilterns. It consists of three parts. In the south-west, between Watlington and Princes Risborough, the cuesta has a classic form, with a straight, high unbroken scarp, and a deeply dissected backslope. In the middle section, between Princes Risborough and Luton, the cuesta is divided by six major gaps, five of which are now dry. In the north-east, the third section has a different form again, especially beyond Hitchin, for the cuesta has a much less classic form. This is probably because this portion was overridden by erosive Pleistocene ice sheets (see p. 53).

East Anglia, the Fens and Bedford lowlands

East Anglia lies to the north of the London Basin and to the south and east of the Fens. As Steers and Mitchell (1967, p. 86) wrote:

> Many think East Anglia is a flat and dull country. But this is not so: at its centre it is true, lies a somewhat featureless plateau but it has lakes, heaths, brecks, fens, river valleys and wide estuaries, and a coast unrivalled for its marshes, its boulder-clay cliffs, and sand and shingle spits. Its geological structure certainly is simple; the rocks are all of Mesozoic and later ages, and none of them offers any serious resistance to denudation.

A glacially scoured Chalk escarpment forms the backbone of the region, but it is a weak feature compared to that which forms the adjoining Chilterns. Most of the Chalk is a low plateau country, mantled in drift and intersected by wide valleys. In the north-west of the region there is a narrow trace of lower Greensand (Lower Cretaceous) which forms a low escarpment facing the Kimmeridge Clay and Fenland. In the south, London Clay and Reading and Thanet Beds outcrop; while in the east there are the later crag deposits.

The imprint of glacial deposition is one of the key features of the area, producing such features as the Cromer Moraine, the outwash plains of Kelling and Salthouse, and the kames and eskers of North Norfolk (see p. 73).

The Fens are one of the most distinctive regions in Britain and their character owes much to their Holocene history. Glacial erosion helped to excavate, between the chalk uplands of Lincolnshire and Norfolk, a broad shallow valley, floored by Jurassic clays, Greensand, Gault and Chalk Marl. In late Glacial times, after retreat of the ice cap, this basin was drained by a system of rivers that were graded to the sea level in a shrunken North Sea. As sea-levels rose in post-glacial times, the North Sea submerged the valley of the Wash, and peat accumulation occurred, burying the bog oaks. At a later date clay deposits – the Buttery Clay – were deposited as the result of a transgression of the sea. These in turn became overlain by peats.

The landscape has been modified more than any other by human intervention, with draining and channelization transforming the hydrology and soils of the area (see p. 308).

In a south-westerly direction the Fens merge into the Bedford lowlands, themselves an extension of the neighbouring Oxford Clay vale. On their eastern side the Bedford lowlands are banded by the Lower Greensand cuesta, which is particularly well developed at Sandy.

The Somerset Plain and south-west England

The Somerset Plain is separated from the Midlands Plain by the hills of the Bristol area and the Mendips. It is underlain for the most part by two main formations: the Keuper Marl of the Trias and Lower Lias. Relief is very low over extensive areas and the Somerset Levels are an area of Holocene alluviation and peat formation. They are markedly different from the landscapes that lie further west.

Most of south-west England, west of Exeter, consists of Palaeozoic rocks, the exceptions being the small areas of Pre-Cambrian igneous and metamorphic rocks which outcrop at the Lizard and Start Point.

The Palaeozoic rocks can be divided into three main groups: Devonian Series shales, slates, sandstone and grits, with some Middle Devonian limestone; the Culm Measures of the lower Carboniferous, also comprising miscellaneous types of sedimentary rock; and the late Carboniferous granite intrusions, with their associated metamorphic aureole.

The sedimentary rocks of the Devonian and the Culm do not in general produce landforms of special interest, though Exmoor, which is composed of Devonian rocks, rises to 520 m. By contrast, the granite bosses of the Late Carboniferous give rise to Dartmoor, Bodmin Moor, Hensbarrow, Carn Menellis and Land's End. They are part of a major granite pluton (batholith) dated at 280 ± 10 million years. The isle of Lundy is part of a much later intrusive phase dating to around 52 million years ago (see p. 27).

Tectonics have played an important role in the evolution of the scenery of the south west, with uplift in the Neogene contributing to the elevated position of Exmoor (Coque-Delhuille, 1987).

The Wessex Downs

To the west of the Weald and inland from the Tertiary lowlands of the Hampshire Basin is an area known as the Wessex Downs. Much of the western portion of this is upland Chalk country, anticlinal in structure. However, the detailed morphology of this area owes much to the existence of numerous subsidiary fold-axes of Tertiary age. Small (1964) has described them thus (p. 40):

> These so-called 'alpine' structures, which are superimposed in the main on to the broad upswelling of the Hampshire–Wiltshire Chalk, are aligned from west to east and increase in amplitude both westward and eastward from a comparatively undisturbed 'structural saddle' near Salisbury and Andover. Most striking of the individual folds are the Warminster and Wardour anticlines of the west, the

Plate 14. Stair Hole, near Lulworth Cove, Dorset, displays the effects of intense folding which took place in the Tertiary as part of the Alpine Orogeny. (Author)

Pewsey anticline of the north-west and the Peasemarsh and Petersfield–Fernhurst anticlines of the east. In the south are some additional and very important folds, including the Weymouth–Purbeck monocline and the Brighstone and Sandown anticlines in the Isle of Wight . . . In cross-section the Alpine folds are normally asymmetrical, with northern dips of up to 30° to 90° and southern dips not exceeding 5°. The steeper northern limbs are in some instances disrupted by strike-faulting as in the western Vale of Wardour . . .

In some cases the anticlinal structures form the ridges and the lowlands are synclines containing Tertiary sediments, but elsewhere (e.g. on the Great Ridge to the north of the Vale of Wardour) 'inverted relief' occurs, with former synclinal valleys left upstanding as hills.

The western portion of the Wessex Downs is developed in a fine succession of Jurassic rocks, including a full succession of Purbeck and Portland Beds, which thin rapidly northwards through England. Indeed the upper Jurassic Portland Beds are a most important scarp former, and they form the tilted slab of the Isle of Portland, with its crest reaching almost 150 m above the eroded Kimmeridge Clay at its foot. The Purbeck Beds, by contrast, are extremely variable in facies and their effect on relief is generally less marked. The Cornbrash and the Corallian tend to form low uplands while the Oxford and Kimmeridge Clays form lowlands. These alternations produce minor scarp and vale relief, notably in the Weymouth lowland (figure 4. 11).

The Middle Jurassic, including the Inferior Oolite, is much less well developed than in the Cotswolds. The Inferior Oolite tends to produce low upland, as does the Forest Marble (limestone and shale) of the Great Oolite, whereas the Fullers' Earth tends to produce vales (Sparks, 1971).

Finally, the Lias has a highly variable lithology and a highly variable effect on relief. The Lower Lias, which consists of alternations of shales and non-massive limestones, tends to produce subdued relief (although there are striking cliffs developed in it near Lyme Regis). The Middle Lias, which tends to be sandier, generally stands out in relief above the Lower Lias and forms a low ridge around the anticlinal Vale of Marshwood. However, a more important scarp former in the area is the Bridport Sands of the Upper Lias.

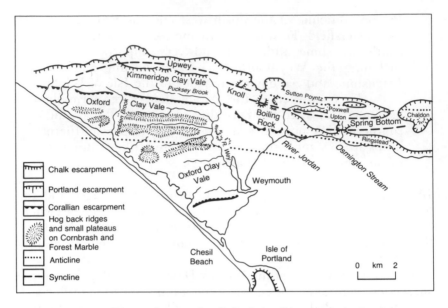

Figure 4.11 The geology and relief of the Weymouth lowland in Dorset (modified from Sparks, 1971, figure 11.7).

The Hampshire and London Basins

Although it has certain similarities to the London Basin, the Hampshire Basin does not have an open-ended fan shape. It is all but closed. It also has a greater degree of asymmetry, and its structural heart lies close to its southern margin. On the northern flank the regional dip is generally less than 3°, whereas on the south the Chalk rim dips deeply and there are narrow hogsback ridges associated with large-scale monoclinal developments in the Isle of Purbeck and the Isle of Wight. The Hampshire Basin's Palaeogene strata are also different, tending to be more continental in facies than are deposits of similar age in the London Basin. There is also the development of Oligocene sediments, which underlie much of the New Forest and the northern portions of the Isle of Wight.

However, as in the London Basin, the relief on the Tertiary Beds is not universally subdued, and gravelly plateau surfaces rise to over 120 m in the northern New Forest.

The London Basin is, in structural terms, an eastward-plunging syncline. It has its apex near Marlborough in Wiltshire and flares out eastwards between diverging chalk rims. The cross-section of the basin is asymmetrical with steeper dips on the southern limb (generally 4–5°) than on the northern (generally less than 2°). Abnormally steep dips occur in the vicinity of the Hog's Back ridge between Fareham and Guildford.

Early Tertiary sands and clays underlie the heart of the London Basin and rest uncomformably on the Chalk. They display considerable variability in lithology. The basal group (Thanet Sands, Reading Beds, Woolwich Beds, and Oldhaven Beds) are predominantly arenaceous and form a marginal strip, abutting on, and sometimes forming outliers on, the Chalk dipslopes. The intermediate group is composed of stiff blue London Clay. By contrast, the upper group, including the Claygate Beds and the Bagshot Beds, create sandy, upstanding heathlands.

The Weald and the Downs

The Weald comprises an oval area in the heart of south-east England (figure 4.12), encircled by the Chalk escarpments of the North and South Downs and the Wessex Downs. Within the bounding Chalk the rocks are sands, sandstones and clays of variable lithology. The great bulk of them is of Cretaceous Age (100–130 million years old). Both they and the Chalk were uplifted and folded into a complex anticline in Late Cretaceous and Tertiary times. The central axis of the dome runs approximately from east to west through the High Weald, incorporating Ashdown Forest and Crowborough Beacon. Beacon Hill rises to 240 m.

Since the Late Cretaceous, erosion has stripped the Chalk off the Wealden anticline exposing the oldest of the rocks in the breached core. The High Weald is underlain by the Hastings Beds, some units of which give rise to valley side cliffs and tors (Robinson and Williams, 1984). The Weald Clay underlies a belt of lowland, called the Low Weald. Looking in towards it is the Lower Greensand escarpment. Where it has a well-developed sequence of resistant Hythe Beds this escarpment can be impressive. At Leith Hill an altitude of 294 m is attained. Beyond the outcrop of the Folkestone Beds at the top of the Lower Greensand, lies a discontinuous Gault Clay vale, followed by the Upper Greensand which sometimes forms a low scarp or bench feature at the foot of the Chalk Downs.

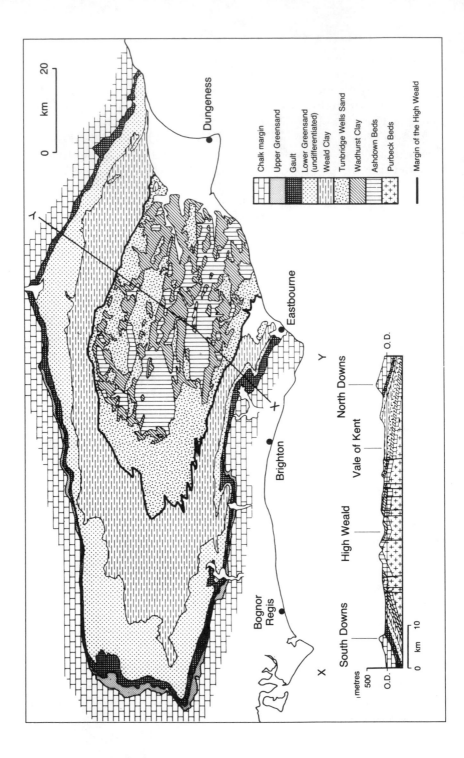

	Chalk margin
	Upper Greensand
	Gault
	Lower Greensand (undifferentiated)
	Weald Clay
	Tunbridge Wells Sand
	Wadhurst Clay
	Ashdown Beds
	Purbeck Beds
—	Margin of the High Weald

Dungeness

Eastbourne

Brighton

Bognor Regis

X

Y

South Downs   High Weald   Vale of Kent   North Downs

metres
500

O.D.

O.D.

0   km   10

0   km   20

Even within the Chalk Downs the effects of lithology are displayed, for the Chalk is far from homogeneous (Jones, 1981, p. 13). The Lower Chalk (often called the Grey Chalk or Chalk Marl) has a high insoluble residue content, most of which is clay, and which normally exceeds 10 per cent but rises to over 50 per cent in lower horizons. The high content of impurities greatly impairs the permeability of this layer, making it prone to subaerial denudation, so that it usually outcrops in a gently inclined footslope at the base of the main scarp. However, within it there are sometimes more resistant beds (e.g. the Totternhoe Stone of the Chilterns). The Middle Chalk is generally purer (insoluble content only 5–10 per cent) and tends to form the main scarp-face beneath a capping of highly permeable and well-jointed Upper Chalk. However, an impersistent nodular layer within the Middle Chalk, the Melbourn Rock, does produce some minor flat-topped promontories on the face of the South Downs and a weakly developed scarp along the north-west margin of the Vale of Pewsey. The Upper Chalk is generally even purer (less than 5 per cent insoluble content), although it is characterized by rows of flints. The lowest 30 m of the Upper Chalk has a resistant, well-jointed horizon called the Chalk Rock; this is the dominant scarp former. However, the *Goniotheuthis quadrata* zone of the Upper Chalk also forms a secondary scarp, but why this is so is not entirely clear. It may be connected with its low insoluble residue content (Sparks, 1949), its seams of nodular flints (Small and Fisher, 1970) its frost resistance (Williams, 1980) or the differential protection beneath a variable cover of Clay-with-Flints (Hodgson et al., 1974).

Chalk may also show variations in mechanical strength according to the degree of dip which the strata display. The study of rock compressive strength and surface hardness with the Schmidt Hammer has shown that, in Dorset, the more steeply dipping strata have greater strength (Jones et al., 1983) as a result of consolidation and pressure solution and recementation. The Needles of the Isle of Wight and the cliffs behind Lulworth Cove are cut in such mechanically hardened Chalk.

Figure 4.12 The geology of the Wealden dome in south-east England with a cross-section from the South Downs across the High Weald to the North Downs (modified from Robinson and Williams, 1984, figure 1).

Structural controls

In addition to the importance of lithology, which has been the main subject of this chapter so far, one has to add the structural history of the crust, for structural trends, developed during past phases of violent mountain building (orogeny) have a clear impact on landscape. Two ancient orogenic events have a particular significance (figure 4.13). The first of these, which occurred between 400 and 500 million years ago, was the Caledonian Orogeny, which imparted a north-east–south-west structural grain of faults and folds across northern Britain. In all, about three-quarters of the total area of the British Isles lies within the Caledonian Fold Belt (Anderson and Owen, 1980), although this is partly hidden by younger formations or modified by younger structures.

The second orogeny took place around 300 million years ago and is called the Hercynian or Variscan orogeny. One of the great orogenic arcs crosses south-west England and is called the Armorican Arc. This orogenic episode imparted an east–west or east–south-east–west-north-west structural pattern across southern Britain, and a fragment of it reaches the surface as the Mendip Hills. It produced lines of weakness which were reactivated by later tectonic episodes. This last point is important because what is called posthumous folding occurs, 'where a cover of relatively young surface rocks experiencing compressional stresses are deformed according to the movements of underlying older and long-buried structures, and take on the orientation of the ancient pattern, irrespective of the direction of the compressive forces' (Jones, 1985, p. 21). Jones elaborates on the significance of this for southern England (p. 21):

> For many years the major folds (the London Basin Syncline, the Weald–Artois Anticline, and the Hampshire–Dieppe Syncline), together with the numerous minor secondary folds superimposed on the main structures (such as the Weymouth, Pewsey, Winchester and Fernhurst folds), were believed to represent the 'outer ripples of the Alpine storm' and to have developed as a result of a brief tectonic episode in the mid-Tertiary. Recent research, however, has shown that such an explanation is untenable. It is now thought that while the creation of the

Alps may have assisted in the development of the young surface fold pattern of southern England, a far greater influence was exerted by the reactivation of Variscan structures, contained in the deeply buried old rocks that underlie the Palaeozoic Floor.

This is not, however, to deny that later tectonic events were unimportant. In the Tertiary, the Alpine Orogeny affected Europe and caused crustal unrest in southern England, where flexures

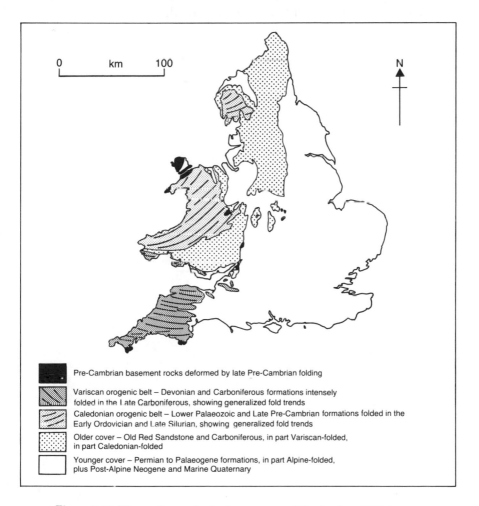

0   km   100

N

Pre-Cambrian basement rocks deformed by late Pre-Cambrian folding

Variscan orogenic belt – Devonian and Carboniferous formations intensely folded in the Late Carboniferous, showing generalized fold trends

Caledonian orogenic belt – Lower Palaeozoic and Late Pre-Cambrian formations folded in the Early Ordovician and Late Silurian, showing generalized fold trends

Older cover – Old Red Sandstone and Carboniferous, in part Variscan-folded, in part Caledonian-folded

Younger cover – Permian to Palaeogene formations, in part Alpine-folded, plus Post-Alpine Neogene and Marine Quaternary

Figure 4.13 The major geological structures of England and Wales.

occurred in the Hampshire and London Basins, Wessex, the Weald, Mendips and the Vale of Glamorgan. Further north, over the Midlands and southern Pennines, broad warping and faulting took place over earlier structural lines and some collapse probably took place at this time in such areas as Cardigan Bay. Faults also cut through the Oligocene beds of the Bovey Tracey Basin in Devon. Northern England and Wales suffered regional upwarping in the Miocene and downwarping occurred in the Irish and North Seas (Anderson and Owen, 1980, p. 25). Such movements probably continued in Post-Miocene times, and subsidence is occurring in the North Sea Basin to this day.

5

Rivers

Introduction

By world standards, the rivers of England and Wales are small and relatively impotent. Some basic data on the largest ones are listed in table 5.1.

Table 5.1 The major rivers of England and Wales

River	Length (km)	Area (km^2)	Mean annual discharge (m^{3}S^{-1})
Thames	239	9,950	67.40
Trent	149	7,490	82.21
Wye	225	4,040	71.41
Tweed	140	4,390	73.85
Severn	206	4,330	62.70
Ouse	117	3,320	40.45
Aire	114	1,930	36.89
Tyne	89	2,180	43.45
Eden	102	1,370	31.02
Ribble	94	1,140	31.72
Great Ouse	184	3,030	14.16
Avon	125	2,210	14.43
Tywi	82	1,910	38.34
Tees	103	1,260	19.46

Source: data in Ward (1981, table 1.1, p. 8)

Given that rainfall is comparatively consistent from month to month and from season to season, river regimes tend to show a peak in winter (figure 5.1), when evaporation is low and groundwater and soil moisture storage are high. They are lowest in the summer months, when evaporation losses are high and groundwater and soil moisture storage are depleted. There is, however, a tendency for the time of both maximum and minimum flows to become gradually later towards the south and east of the country. The mean maximum monthly flow in north-west England and in the western parts of Wales tends to occur in December, but it is as late as February and March in some eastern areas. The mean minimum monthly flow occurs in May in some Cheshire rivers, June in northern England and parts of northern and western Wales, but as late as September in parts of eastern or central southern England (Ward, 1981).

It is therefore probable that a substantial proportion of the geomorphological work achieved by the rivers of England and Wales takes place in the winter months. However, as we shall see, many major flood events of geomorphological significance occur in the summer months as a result of convective rainfall. There is also some evidence that the incidence of geomorphologically significant flooding has increased, in some cases because of human actions, but in others because of natural climatic change.

Extreme events

Although normal levels of rainfall intensity in Britain are low, there are occasions when high-magnitude rainfall events occur and prove to be capable of causing geomorphological changes of some considerable significance. Such changes include gullying, delta and fan deposition, channel scour, cave sediment flushing, debris flows, and peat bursts (Newson, 1980). Antecedent moisture conditions are important but, bearing that in mind, there appears to be a threshold of about 50 mm per hour required for major flood events to occur. Data and dates for major flood events are shown in table 5.2. The summer months are plainly the most important ones for intense storms with August being the dominant month, accounting for over one-third of all notable floods (Newson, 1975) because of the development of intense, thundery, convective storms.

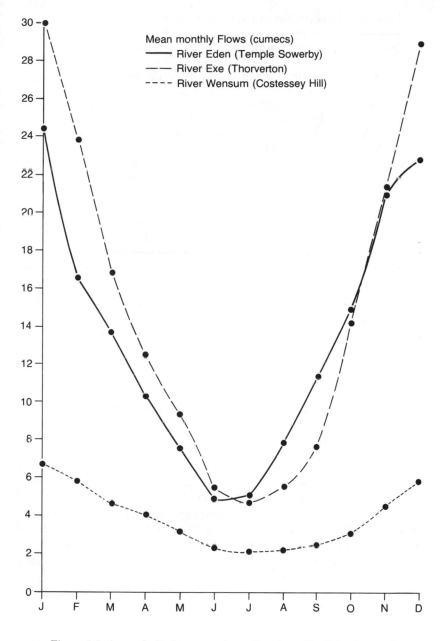

Figure 5.1 Annual discharge regimes for three English rivers: the Eden, Exe and Wensum.

Table 5.2 Twentieth-century British floods

Date	Location	Rainfall
29.5.20	Louth, Lincolnshire	153 mm 3 h^{-1}
17.5.36	Chilterns	42 mm
29.5.44	North–mid-Wales	54 mm 1.5 h^{-1}
15.8.52	Lynmouth, Devon	>225 mm 24 h^{-1}
18.7.55	Weymouth, Dorset	279 mm 6–9 h^{-1}
8.6.57	Camelford, Cornwall	138 mm 2.5 h^{-1}
6.8.57	W Derbyshire	150 mm 5 h^{-1}
8.8.67	Forest of Bowland	117 mm 1.5 h^{-1}
10.7.68	Mendip	101 mm 2 h^{-1}
5.8.73	Mid-Wales	72.7 mm 6 h^{-1}
15.8.77	Mid-Wales	86 mm 1.3 h^{-1}
14.8.75	London	171 mm 3 h^{-1}

Source: Newson (1975, 1980)

There were 151 occasions in the period 1863–1970 when 125 mm or more precipitation occurred during a day in Britain (Bleasdale, 1970) and, while most of these events occurred in upland areas like the Lake District, Snowdonia, South Wales and the moors of south-west England, the most intense events occurred in Somerset and Dorset:

18.7.55.	Martinstown, Dorset	297.4 mm/day
28.6.17.	Bruton, Somerset	242.8 mm/day
18.8.24	Cannington, Somerset	238.8 mm/day

A good review of British rainfall excesses is given by Rodda (1970).

There is some evidence that serious flood frequencies have shown some increases in recent decades. Howe et al. (1966) analysed flood data for the River Severn (at Shrewsbury) and the Wye (at Hereford). They found that during the period 1911–40 a flood height of 5.1 m was to be expected at Shrewsbury only once in 25 years; during the period 1940–64 the Severn reached this height once every four years. Figure 5.2A shows the trend for the Wye, and demonstrates the great clustering of high flood events since about 1930–40. The reasons for these increasing flood levels are complex, but probably include deliberate peat drainage in the Welsh uplands and an increase in the frequency of daily rainfalls greater than 63.5 mm since 1940 (figure 5.2B). Rodda's work (1970; figure 5.2C) showed that an increase in intense rainstorms had also occurred in central England (as represented by climatic data from the Radcliffe Meteorological Station, Oxford). For the period 1881–1905 the return period for a storm of just over 50 mm was about 30 years, but for the period 1941–65 the return period of the same size of daily fall had dropped to little more than five years.

This climatic explanation for increasing flood levels has been confirmed for South Wales by Walsh et al. (1982). In the case of the Tawe valley near Swansea, of 17 major floods since 1875, 14 occurred from 1929–81 and only three during the 1875–1928 period. Significantly, of 22 notable widespread heavy rainfalls in the Tawe catchment since 1875, only two occurred from 1875 to 1928, but 20 from 1929 to 1981. However, Lawler (1987) has sounded a note of caution about overgeneralizing about this particular climatic trend and argues not only that no simple synchroneity across the country exists in any changes identified but that since 1968 there has in some areas been a reversal in the trend of increasing storm rainfalls and associated floods.

Run-off processes

The classic model of run-off generation is that of R. E. Horton (1945), who proposed that when rainfall intensity is greater than soil infiltration capacity, overland flow ('infiltration excess overland flow') will occur widely over a catchment area. However, this American model seems of limited applicability in the British context, where rainfall intensities are generally low, soils have good structures which promote high infiltration rates (table 5.3) and there is a cover of vegetation and soil litter to protect the soil surface from raindrop sealing and crusting.

It is likely that a more applicable model is the through-flow model of Kirkby and Chorley (1967). In this model it is proposed that the permeability of the soil tends to decline with depth and that this places a limit upon the amount of deep infiltration that can occur. Thus, within the soil, above an horizon of reduced permeability, lateral through-flow occurs. Soil moisture content thus increases downslope, but only tends to approach saturation in a zone adjacent to a stream channel. Overland flow only occurs over the lower portions of a hill-slope where a saturated surface is present. Additional run-off will be created by direct precipitation on to channels. Thus, under this partial area model, only a limited

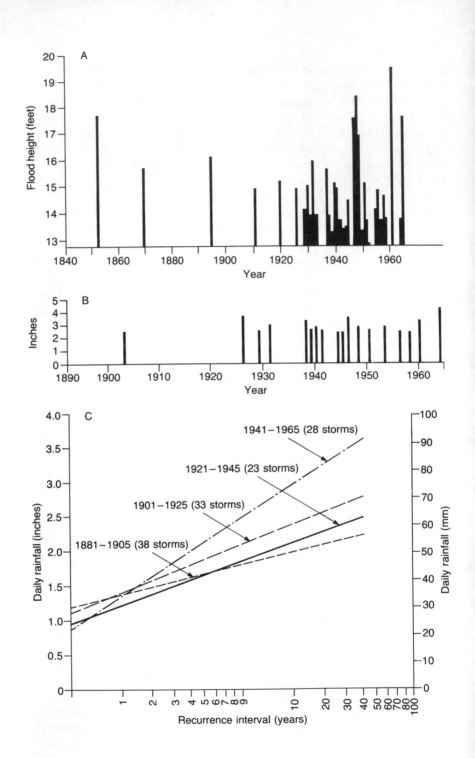

Table 5.3 Infiltration rates in the Bristol area

Land use	Soil series	Infiltration capacity (mm h^{-1})	
		Winter	Summer
Bare, compacted orchard	Worcester	0.5	9
Woodland	Worcester	50	—
Light pasture	Nibley	26	115
Woodland	Bank	191	183
Light pasture	Bank	68	366

Source: data in Hills (1971)

portion of the basin will be a contributing area to run-off. Thus in his analysis of two small Mendip catchments Weyman (1974) found that the run-off contributing areas were up to 32 per cent of basin area in one case, but only 0.7–2.0 per cent in another. The differences were due to concave contours and a peaty-gleyed podsoil in the first case, and to convex contours and free-draining brown earth in the second.

Lateral movement of soil water may be accelerated by the presence of soil pipes and percolines (Bunting, 1964). These occur widely in Britain and have for example, been noted in the Mendips, the Peak District, the Plynlimon catchments of Wales, Berkshire, the Brecon Beacons, the Lake District, and the Isle of Wight (Jones, 1981).

Gilman and Newson (1980, p. 99) remark that, 'Geomorphologically the importance of soil piping on hill slopes is that along with soil creep, it is a process which can work effectively on Britain's "arthritic concavo-convex hillslopes, crippled and reduced to impotence beneath a blanket of sod" (King, 1953)'.

Pipes are small, subsurface tunnels formed by processes of suffosion. They come in a variety of forms and Gilman and Newson (1980) draw a distinction in the catchment of the upper Wye

Figure 5.2 Changes in flood levels and daily rainfall amounts: (A) recorded flood levels of the River Wye at Wye Bridge, Hereford; (B) the frequency of daily rainfalls of greater than 2.5 inches (c.64 mm) at Lake Vyrnwy, mid-Wales; (C) changes in the magnitude–frequency relations for daily rainfall amounts exceeding 25 mm at Oxford (after Rodda, 1970, figure 8).

between deep-seated large pipes flowing seasonally or perenially, and the smaller hill-slope pipes close to the surface and flowing ephemerally. The former, which have a diameter greater than 200 mm, may, they believe, be considered as peat-bridged streams, occupying the base of drift-filled channels. From time to time the peat roofs may collapse to produce vertical shafts and pipe blockage. This type of pipe is less common than the second type, which have a smaller diameter (c.40 mm), 'are unassociated with obvious physiographic features, drain areas of the catchment flanks, and flow ephemerally in response to rainstorms' (Gilman and Newson, p. 25). Ephemeral pipe systems may have a very dense network, and densities in the Wye catchment reached 180 km km^{-2}.

Percolines were identified in the Matlock area of Derbyshire by Bunting (1964). These are defined as 'lines of over-deepened soil' along which moisture accumulates and moves laterally within the soil. In his field area they have widths of 30–100 m, attain lengths of 600 m, and have deep soils (90–100 cm) compared with those of interpercoline areas (60–80 cm).

The importance of through-flow processes in small British catchments has been established by empirical studies of hydrographs. For example, in his study of discharge from the Preston Montfort Brook in Shropshire, Job (1982) found that high-intensity rainfall events produced two hydrograph peaks: the first, which occurred with a lag time of about one hour from the main rainfall input reflected rapid run-off from direct channel precipitation, roads and possibly a pond; the second, which occurred with a lag of at least five hours, was related to slower through-flow processes assisted by the presence of tile drains. Such double peaks have also been found to occur after some storms in the Slapton Wood catchment in South Devon (Burt et al., 1983a) though whether they do or not depends on the intensity of the rainfall and on the nature of the antecedent moisture conditions. The lag time to the second peak in Slapton can be of the order of two days.

Solute loads and chemical denudation

Specific conductance gives an indication of the total dissolved solids concentration of stream water, and Walling and Webb (1981) have attempted to map its pattern (figure 5.3A). The crucial importance of rock type (and associated elevation) is clear. Upland areas developed on resistant rocks (the Lake District, Snowdonia, mid-Wales and parts of the south-west peninsula) have generally low specific conductance values, whereas lowland areas to the east of the Tees–Exe line, developed on less resistant rocks (especially sedimentaries) have much higher levels. Characteristic specific conductance levels in streams draining major rock types are shown in table 5.4. This pattern reflects the general dilution effect exercised by high run-off from moist upland areas and the susceptibility of sedimentary rocks to chemical attack.

Table 5.4 Characteristic specific conductance values in streams draining major rock types

Rock type	Conductance values ($\mu S\ cm^{-1}$)
Granite and Metamorphics	$<$ 40–90
Resistant Palaeozoic sandstones, slates and shales	40–300
Carboniferous Limestone	200–500
Permo-Triassic, Mesozoic, and Tertiary sedimentaries	$>$ 300
Chalk	400–800
Boulder Clay over Mesozoic sediments	$>$ 700

Source: Walling and Webb (1981, table 5.7., p. 137)

From this one can look at the pattern of total dissolved solids loadings expressed as the rate of removal in tonnes $km^2\ y^{-1}$ (figure 5.3B). This is based on an analysis of solute concentrations and run-off volumes. Values range between 10 and 400 tonnes $km^2\ y^{-1}$. Areas in the east of England, being drier, have low or medium dissolved-solid loadings as do upland areas with highly resistant rocks.

In areas of resistant rock quite substantial proportions of stream solute may be derived from inputs in precipitation, with sodium and chloride being the dominant ions (table 5.5). This is especially true if allowance is made for the concentration effect caused by evaporation loss. Indeed, in some cases more ions may come in than are evacuated in stream water. It is instructive to compare data on rainfall and stream chemistry for the Upper Severn in mid-Wales (table 5.6).

Thus, to calculate rates of chemical denudation it is necessary to apportion solute loads in denudational and non-denudational components. When this is done, it is found that in general rates of solutional denudation on non-carbonate rocks (for comparative data on carbonate rocks see table 7.1) are low (table 5.7) with most values being equivalent to a denudation rate of less than $40 \text{ m}^3 \text{ km}^{-2} \text{ y}^{-1}$. Walling and Webb (1986) have produced a map of chemical denudation rates for Britain (figure 5.4), based on 1,600 sites 'located on unpolluted streams of small and moderate size', using available information on discharge and water chemistry, and subtracting the non-denudational component. They calculate that more than 50 per cent of the country experiences an annual rate which is less than $20 \text{ m}^3 \text{ km}^{-2} \text{ y}^{-1}$.

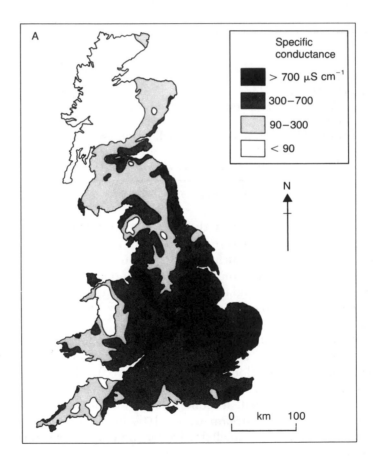

Figure 5.3 Solute loads of British rivers: (A) mean background loads of specific conductance in streams, based primarily on data collected

Suspended and bed-load sediment transport

The quality and quantity of data on suspended sediment transport
in the rivers of England and Wales is poor, and the problem is
compounded by the use of different measurement and sampling
strategies by different workers (Walling and Webb, 1981). Furthermore, in comparison with solute loads, river basin area is probably
a more important influence on suspended sediment loadings
because of the different possibilities for storage within catchments
of different sizes.

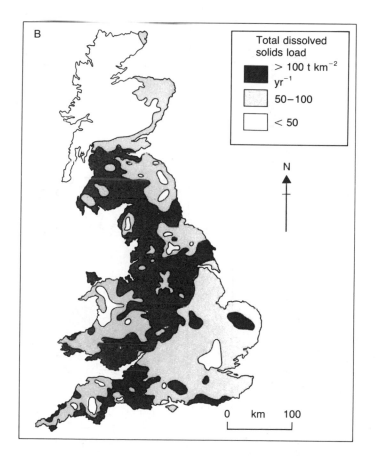

in the period 1977–9; (B) the pattern of annual total dissolved solids
loads in streams (from Walling and Webb, 1981, figures 5.3 and 5.9).

Table 5.5a British precipitation inputs (kg ha^{-1} y^{-1})

Location and rainfall	Source	Ions					
		N	P	Na	K	Ca	Mg
Moor House, Peak District (186.05 cm)	Gore (1968)	14.04	0.85	35.06	4.80	13.10	3.23
Abbot's Moss, Cheshire (95 cm)	Allen et al. (1968)	14.0	0.8	14	5.4	14.0	2.9
Grizedale, N Lancs (169 cm)	Allen et al. (1968)	7.6	0.3	44.3	3.4	12.4	5.4
Merlewood, N Lancs (150 cm)	Allen et al. (1968)	15.0	1.0	22	3.7	12.0	3.6
Silpho, N Yorks (116 cm)	Allen et al. (1968)	13.0	0.8	36	4.8	9.8	4.5
Ascot, Berks (88 cm)	Alcock and Morton (1985)	12.9	1.2	—	2.5	34.4	2.0
Mean		13.6	0.8	30.3	4.1	16.0	3.6

Table 5.5b British precipitation quality (ppm or mg l^{-1})

Location	Source	Ions					
		Ca	Mg	Na	K	Cl	SO$_4$
Plynlimon, mid-Wales	Lewin et al. (1974)	1.49	0.38	2.60	0.17	4.84	—
Plynlimon, mid-Wales	Neal et al. (1986)	0.21	0.41	3.24	0.5	6.08	1.85
Devon, England	Foster (1979)	2.4	0.9	3.7	1.9	6.8	—
Leeds, Yorks	Stevenson (1968)	2.7	1.0	2.3	—	5.3	—
Camborne, Cornwall	Stevenson (1968)	0.8	0.6	7.7	—	11.3	—
Rothamstead	Stevenson (1968)	1.4	0.3	1.8	—	3.1	—
Newton Abbot, Devon	Stevenson (1968)	1.5	0.6	5.5	—	6.3	—
Ascot, Berkshire	Alcock and Morton (1985)	3.9	0.22	—	0.28	—	—
Mean		1.8	0.6	3.8	0.6	6.2	—

Table 5.6 Rainfall and stream chemistry in the upper
River Severn catchment, Wales (Plynlimon), expressed
as mean concentrations $(mg\ l^{-1})$

Ion	Rainfall	Hafren Stream	Hore Stream
Na	3.24	4.37	4.53
K	0.15	0.20	0.15
Ca	0.21	0.81	1.15
Mg	0.41	0.80	0.82
SO4	1.85	4.53	4.93
NO3	0.80	1.88	1.54
Cl	6.08	8.56	9.29

Source: Neal et al. (1986)

There is a very large range in suspended sediment yields of
British rivers, with a spread from less than 1.0 tonnes $km^{-2}\ y^{-1}$
to almost 500 tonnes $km^{-2}\ y^{-1}$. Walling and Webb (1981) regard
a value of 50 tonnes $km^{-2}\ y^{-1}$ as typical of British rivers, and
regard this as low by global standards. This is probably because of
the low peak intensities of rainfall and the relatively dense
vegetation and crop cover (Moore and Newson, 1986).

Estimates of long-term sediment yield based on reservoir surveys
are few in number but give figures of the same order of magnitude
as the river sediment yield data based on sediment load data
discussed above. The average rate of sediment supply to reservoirs
is approximately 30 tonnes $km^{-2}\ y^{-1}$.

These two types of method for estimating sediment yield are too
sparse to enable a map comparable to figure 5.4 to be constructed.
However, Walling and Webb (1981, p. 167) allow themselves a few
tentative generalizations:

Loads in excess of 100 y $km^{-2}\ y^{-1}$ would seem to be associated primarily with upland areas receiving annual precipitation greater than 1000 mm and, within these areas,
with small- and intermediate-sized catchments where
sediment delivery ratios will be relatively high. Conversely
low suspended sediment yields (< 25 t $km^{-2}\ y^{-1}$) would
appear to reflect low annual precipitation (e.g. River
Welland and River Nene), large basins where sediment
delivery ratios will be relatively low . . . and low relief . . .
Very low suspended sediment yields (< 5 t $km^{-2}\ y^{-1}$),

Table 5.7 Rates of solutional denudation on non-carbonate rocks

Lithology	Location	Solutional denudation (mm 1000 y^{-1})	
Old Red Sandstone	Mendip	1.6	Waylen (1979)
Clay-with-flints	SE Devon	14.0	Walling (1971)
Keuper Marl	SE Devon	22.8	Walling (1971)
Upper Greensand	SE Devon	16.8	Walling (1971)
Keuper Marl–Upper Greensand	SE Devon	33.8	Walling (1971)
Keuper Marl–Upper Greensand	SE Devon	27.6	Walling (1971)
Silurian Greywackes	Montgomeryshire	2.1	Oxley (1974)
Silurian Greywackes	Montgomeryshire	1.6	Oxley (1974)
Middle Jurassic–Liassic clays and shales	North York Moors	20.2	Imeson (1974)
Granite	Dartmoor	5.0	Williams et al. (1986)
Sandstones and shales	Midlands	10.5	Foster and Grieve (1984)

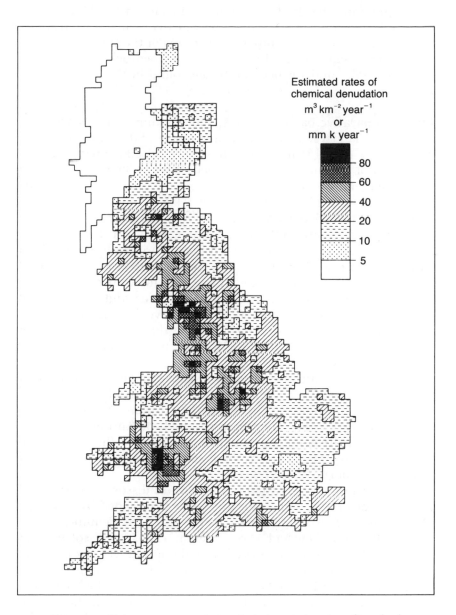

Figure 5.4 Estimated rates of chemical denudation in $m^3 km^{-2}y^{-1}$ (mm 1000 year^{-1}), (after Walling and Webb, 1986, figure 7.18).

represented by the East Twin catchment on the Mendip Hills and the Ebyr N. and Ebyr S. catchments in central Wales, may be accounted for in terms of the small headwater areas involved, the resistant bedrock and the essentially undisturbed conditions found in these upland areas.

On a comparative basis, river suspended-sediment loads are often rather less than river solute yields, and values of the sediment:solute ratio greater than 1.0 are for the most part associated with uplands composed of chemically resistant rocks. Out of 35 rivers analysed by Walling and Webb (1981, table 5.13) only nine (i.e. 26 per cent) had ratios > 1.0. The mean ratio was 0.92.

Not all suspended load transport is necessarily composed of mineral matter derived from the mineral soil or bedrock of a catchment area. Streams may transport organic material in suspension and such material may be derived from a variety of sources, including invertebrate fauna, unicellular and filamentous algae and phytoplankton, invertebrate exuviae and fragments of dead plant and animal material. Quantitative data are sparse, but Finlayson's work in the Mendips indicates that the organic fraction of suspended solids in streamflow may be around 20 per cent (Finlayson, 1978).

Data on bed-load transport are even sparser than on suspended sediment transport. This reflects acute problems of measurement and is not a phenomenon restricted to Britain. Newson (1981, p. 77) suggests that, 'the high residual proportion of coarse materials in British upland channels means that the majority of sediment transport occurs as bedload', whereas in non-upland rivers the proportions are reversed.

A rare example of an attempt to measure dissolved, suspended and bed loads at the same time and place is provided by Carling's (1983) study of two catchments in the northern Pennines (table 5.8a). Although they had very different ratios of dissolved and suspended loads, bed load was of limited importance in both cases. The same is not true for Reynolds's study in the headwater catchments of the Upper Wye, in mid-Wales, where bed load was more significant than suspended load (table 5.8b). In the case of two Shropshire catchments, bed load was insignificant and dissolved load was more important than suspended load (table 5.8c, d).

Table 5.8 Measurements of sediment load

Location and Load	Quantity (t km^{-2} y^{-1})	Percentage
N Pennine basins[a]		
Carl Beck		
Dissolved	110.28	81.5
Suspended	24.77	18.3
Bed	0.33	0.2
Total	135.38	100.0
Great Eggleshape Beck		
Dissolved	0.00	0.0
Suspended	12.07	95.6
Bed	0.55	4.4
Total	12.62	100.0
Afon Cyff, Upper Wye, mid-Wales[b]		
Dissolved	37.6	77.7
Suspended	2.8	5.8
Bedload	8.0	16.5
Total	48.4	100.0
Shropshire catchments		
Preston Montfort[c]		
Dissolved	67–100	—
Suspended	32.8	—
Bedload	1.05	—
Total	100.85–123.85	—
Farlow[d]		
Dissolved	113.02	50.219
Suspended	112.03	49.780
Bedload	0.0002	0.001
Total	225.052	100.00

[a] From Carling (1983)
[b] From Reynolds (1986)
[c] After Job (1982)
[d] After Mitchell and Gerrard (1987)

Rates of channel change: river-bank erosion

Rates of river-bank erosion can be determined either by direct measurement or by using sequences of maps. Hooke (1980) and Newson (1986) provide data on rates for England and Wales; these are listed in table 5.9. Several things can be said about these rates. First, bank erosion rate, as might be expected, is closely related to catchment area (a surrogate of discharge and width). Secondly, rates of bank erosion are such that this may well be an important source of stream suspended loads. Thirdly, because river channel movement is one of the main processes by which the formation, development and renewal of floodplains takes place, these data have important implications for rates for floodplain development. Hooke calculates that, for rivers in Devon (using measured flood plain widths and mean rates of flood plain development), it takes between around 6,000 and 7,000 years for a complete traverse of the floodplain to take place. Fourthly, the data indicate that at some sites rates are sufficiently high as to cause considerable practical problems for riparian owners.

However, as Ferguson (1981) points out, most British rivers have rather inactive channels and have shown little change since their courses were surveyed accurately for the first time in the late nineteenth century. Some rivers are constrained in rock-walled

Table 5.9 Rates of river bank erosion

River	Dates of surveys	Rate (m y^{-1})
Cound, Shropshire	1972–4	0.64
Bollin, Cheshire	1872–1935	0.16
Exe, Devon	1840–1975	2.58
Creedy, Devon	1840–1975	0.52
Culm, Devon	1840–1975	0.51
Axe, Devon	1840–1975	1.00
Yarty, Devon	1840–1975	1.38
Coly, Devon	1840–1975	0.48
Hookamoor, Devon	1840–1975	0.19
Severn	1948–75	0.2–0.7
Rheidol	1951–71	1.75
Tywi	1905–71	2.65

Source: Newson (1986, table II); Hooke, (1980, table IV)

Plate 15. Durham Cathedral dominates the City of Durham and lies on a meander core of the River Wear. (Michael J. Stead)

channels (e.g. the Wye in the Welsh borderland, and the Dove and the Wye in Derbyshire). Others have banks which are protected against erosion by the presence of trees. However, the biggest class of inactive channel in Britain is the clay lowland rivers of East Anglia and the Thames Basin, where low regional slopes, a small rainfall surplus over evaporation and cohesive banks, derived from soft bedrock or Pleistocene drifts, combine to dampen down present-day channel migration.

Although channel changes are ultimately caused by river erosion and transport, detailed monitoring of meander beds in South Wales by Lawler (1986), using grid networks of erosion pins, has highlighted the importance of frost action in preparing banks for subsequent fluvial scour at times of high flow. He found that minimal fluvial erosion occurred on banks which had not previously been prepared by frost attack and that most bank retreat occurred in the winter months of December, January and February.

River longitudinal profiles and gradients

The only comprehensive analysis of the shape of river long profiles in Britain is that of Wheeler (1979). He used 1:25,000 OS maps to determine the profiles of 115 rivers and then calculated an index of concavity based on the maximum departure below a constant gradient of the percentage-drop–percentage-distance profile. Concavities showed a large range, from 0.09 for the Colne in Hertfordshire, to 0.72 for the Irt in Cumbria. There is also a general relationship between stream relief and concavity such that basins with greater available relief tend to have higher indices of concavity (figure 5.5).

However, there are some atypical profiles which lie away from the line of the regression equation expressing this relationship: they are either 'overconcave' or 'underconcave'.

The two clearest examples of overconcavity are the Yorkshire Derwent and the Somerset Parrett. The former rises at 261 m on the eastern flanks of the North York Moors and falls to an elevation of only 25 m in the first 27 km of its 112 km course. For much of its course it flows over the lowlands of the Vale of York with its extensive infilling of glacial deposits. The Parrett rises at 147 m on the Dorset Heights and reaches 25 m within 13 km of its 64 km course. It then follows a course over the Somerset Levels, an area of

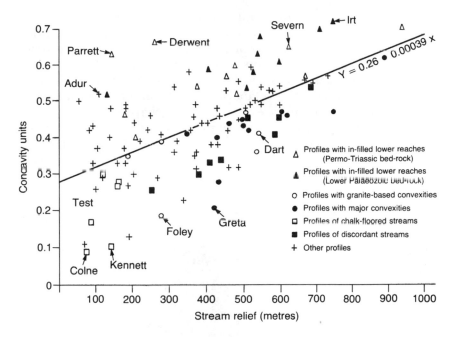

Figure 5.5 The association of stream relief with overall concavity in stream long profiles (after Wheeler, 1979, figure 14.3).

Holocene alluvial infilling (see p. 143). Thus overconcavity seems to be a consequence of relatively recent infilling of a river's lower reaches by glacial and post-glacial deposits. Indeed, most streams ending in major estuarine lowlands have a tendency of this type. A second category of overconcave streams identified by Wheeler is that affected either by glacial aggradation or that affected by glacial diversion and subsequent lengthening of courses (e.g. the Severn, see p. 517).

There are two types of underconcave river: those caused by geological factors, and those caused by abnormal patterns of discharge change downstream. The first of these is caused by the presence of marked structures such as faults across a river's course or by abrupt lithological contrasts along their profiles. The Greta in north-west Yorkshire has marked underconvexity because it has to contend with the Craven Fault system; while in south-west England rivers like the Dart and Teign and the Fowey steepen their gradients where they leave the granite outcrops of Dartmoor and Bodmin Moor respectively.

These geologically caused underconcavities incorporate a major convexity within their courses, but there are some rivers which have profiles that are less concave simply because their profiles overall are relatively constant in gradient. This is the case with some of the rivers that traverse the chalklands of southern England, such as the Kennet. The explanation may be in the abnormally low downstream rate of increase in their discharges. This in turn reflects the relatively low rainfall of the area, the permeability of the chalk and the associated predominance of groundwater discharge.

River long profiles with the same degree of concavity may have very different overall gradients (Ferguson, 1981). Figure 5.6 gives a

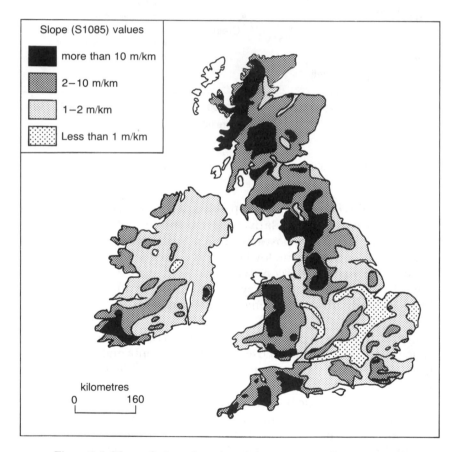

Figure 5.6 Map of river-slope based on mean gradients between points 10 and 85 per cent along individual rivers (after Newson, 1978, figure 4.6).

regional picture of river slope, and the map is based on mean gradients between points 10 per cent and 85 per cent along individual rivers. It is apparent, as one might expect, that rivers rising in the more mountainous north and west have steeper gradients. There are, however, substantial local variations in gradient on account of geological and hydrological differences and because of glaciation. Furthermore, differences in slope exist because rivers vary in size: gradient decreases with the length of river considered (Penning-Rowsell and Townshend, 1978).

Drainage and network densities

There have been few attempts to provide a nationwide picture of drainage basin characteristics. An exception is the work of Gregory (1976) who determined drainage densities for ten sample basins in 15 areas of Britain, using the delimitation of stream networks shown on 1:25,000 Provisional Edition OS maps. He found that when the average densities from each area were considered in relation to mean annual precipitation totals, there was a general association of increase in drainage density with increased amounts of precipitation. He also grouped the 15 areas according to whether permeable or impermeable rock type prevailed and found, as might be expected, that the impermeable catchments had the higher drainage densities.

In all cases the drainage densities were low. The mean value for all 15 areas was 1.57 km km^{-2}, and the range was from 0.95 to 2.28 km km^{-2}. However, under flood conditions, drainage densities may in reality be higher than these map-derived figures suggest, and studies of individual basins during storm events indicate that values may be as high as 4 to 6 km km^{-2} (Gregory and Walling, 1968).

However, it is normal, as the study of dry valleys suggests (see p. 176) for valley densities to be generally higher than stream densities (table 5.10). Even so, values are still low on a global basis, indicating the modest development of drainage incision into the British landscape.

The nature of drainage networks has been considerably modified in recent decades by the installation of field drains and many catchments now have a substantial proportion of their area drained (figure 5.7). The effects of such drainage operations on flood run-off

Table 5.10 Valley densities for English catchments

Source	Location	Mean density (km km^{-2})
Brunsden (1968)	Dartmoor granite	1.61–2.24
	Middle Devonian	2.92–2.98
	Lower Devonian	2.86
	Culm measures	3.23–4.78
Morgan (1971)	Weald Clay	2.3
	Ashdown Sand	2.0
	Tunbridge Wells Sand	1.7
	Chalk	2.3
Gregory (1976)	Jurassic Limestone (Yorks)	2.32
	Chalk (Yorks)	1.14
	Carboniferous Limestone (Peak District)	1.72
	Chalk (Lincolnshire)	1.51
	Chalk (Hampshire)	1.89
Dalton and Fox (1986)	Triassic sandstone (Derbyshire)	2.81–3.25

will vary according to such factors as the intensity of the drainage works (depth, spacing, etc.), and the nature of the soils being drained (Robinson, 1979). These may be two (sometimes conflicting) processes operating as a result of drainage: the increased drainage network will facilitate rapid run-off, whereas the drier soil conditions will reduce the amount of saturation excess overland flow and provide greater storage for rainfall.

Dry valleys

One of the most widespread and most discussed features of the British landscape is the dry-valley network that is developed on many rock types. The classic localities are on Chalk, and it was for these that Reid (1887, p. 370) put forward his famous periglacial theory of origin, in which he invoked the important role of permafrost:

All rocks would be equally and entirely pervious to water, and all springs would fail. While these conditions lasted, any rain falling in the summer would be unable to penetrate more than a few inches. Instead of sinking in the Chalk, or other pervious rock, and being slowly given out in springs, the whole rainfall would immediately run off any steep slopes like those of the Downs, and form violent and transitory mountain-torrents. These would tear up a layer of rubble previously loosened by the frost and unprotected by vegetation.

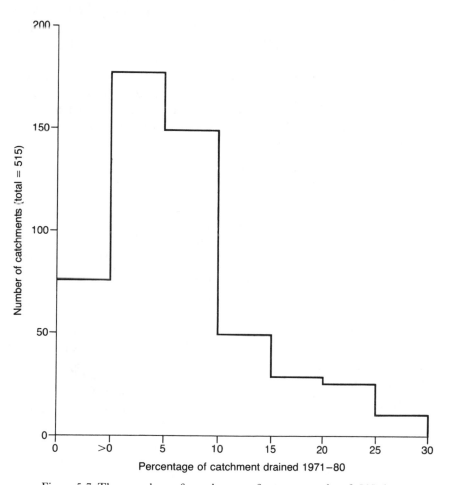

Figure 5.7 The number of catchments from a sample of 515 in England and Wales with a given percentage of their area drained, 1897–1980 (from Institute of Hydrology Research Report, 1984–1987, figure 4.6)

However, dry valleys occur on a whole range of rock types besides the Chalk, including the Triassic sandstone outcrops of south Derbyshire (Jones, 1979; Dalton and Fox, 1986); the Triassic beds of east Devon (Gregory, 1971); the Triassic sediments of Cannock Chase; Quaternary fluviglacial sands and gravels in north Norfolk (Richards and Anderson, 1978); the Carboniferous limestones of the Pennines (Warwick, 1963); the Jurassic limestones of the Cotswolds (Beckinsale 1970); and the Arden Sandstones (Triassic) of Warwickshire (Jones, 1968).

There are a vast number of hypotheses that have been developed to explain the presence of dry valleys. These can be categorized into uniformitarian hypotheses (those requiring no major changes of climate or base level, but merely the operation of normal processes through time); into marine hypotheses (those related to base-level changes); and into palaeoclimatic hypotheses (those associated primarily with the major climatic changes on the Pleistocene). These hypotheses are listed in table 5.11. It is therefore clear that Reid's hypothesis is only one of many possible explanations.

Table 5.11 Hypotheses of dry valley formation

Uniformitarian	Marine	Palaeoclimatic
Superimposition from a cover of impermeable rocks or sediments	Failure of streams to adjust to a falling Pleistocene sea-level and associated fall of groundwater levels	Overflow from proglacial lakes
Joint enlargement by solution over time (i.e. increased secondary permeability)	Exposure of former submarine canyons	Erosion by glacial meltwaters
		Glacial scour
Cutting down of major through-flowing streams (leading to a reduction in altitude of water table)	Tidal scour in association with former estuarine conditions (tidal palaeomorphs)	Spring snowmelt under periglacial conditions
		Run-off from impermeable permafrost
Fall of groundwater levels arising from scarp retreat (Chandler-Fagg hypothesis)		Reduced evaporation caused by lower temperatures
Reduction of catchment area following scarp retreat or river capture		Increased precipitation
Cavern collapse		
Rare events of extreme magnitude		

The uniformitarian hypotheses cannot be dismissed. Some dry valleys appear to head back into areas of impermeable rocks that overlie rocks like limestone, and in such cases it is entirely possible that the former were superimposed from an impermeable stratum that has since been removed by erosion. Warwick (1963) viewed the dry valleys of the South Pennines as having been inherited from an impervious Namurian cover. Likewise, it is apparent that in a limestone area, as time goes on, the joints in the limestone will be progressively enlarged by solution, thereby reducing the volume of surface stream flow (Smith, 1975c). In the same way, over time a master stream may cut down preferentially to the others and thereby abstract some of the water from their basins, thus making them go dry. The retreat of an escarpment through time is an indisputable fact (revealed, for example, by the breaching of the Wealden Dome or the presence of outliers detached from the main Cotswold escarpment) and this is another factor that could reduce the catchment area of a stream and a reduction of groundwater levels (Chandler, 1909; Fagg, 1939). The collapse of underground caverns has been evidenced to explain large dry valleys like Cheddar (see Ford and Stanton, 1968, for a critique of this view) and the capture of one river by another could cause localized examples of dry valleys, although it could scarcely account for a widespread pattern. It is also important to remember that extreme rainfall events of low frequency can sometimes cause so-called dry valleys to flow. This happened in Cheddar Gorge in the Mendips in July 1968 (Hanwell and Newson, 1970).

The recognition that sea-levels have changed during the course of the Pleistocene and that in Tertiary times certain parts of Britain were submerged by the sea (e.g. the proposed Calabrian transgression to around 180 m in parts of south-east England) has tempted some authors to interpret dry valleys and misfit streams in terms of base-level change. At times of high sea-level groundwater levels would also have been high but, as the sea retreated, the groundwater level would also have fallen, causing some of the valleys to go dry. It is also possible that some valley systems might have been created by tidal scour (Geyl, 1976) or by turbidity currents (Winslow, 1966), but the supporting evidence is largely speculative.

The most widely adopted explanations for dry valleys are those involving palaeoclimatic changes associated with rigorous glacial or periglacial climates in the Pleistocene. Glacial agency may explain some dry valleys within the areas subjected to the direct effects of glaciation such as parts of the Yorkshire Wolds (Foster,

1987), but of more general significance is the effect of periglaciation. Cold climatic conditions would have caused depressed evaporation rates so that run-off would be a greater proportion of precipitation. Much of the precipitation would have fallen in the winter months as snow and, when this melted in the summer thaw period, rapid run-off would have occurred (the nival hypothesis of Beckinsale, 1970). The presence of an impermeable permafrost layer, as proposed by Reid, would then have enabled surface flow on rocks that are normally too pervious to permit it (e.g. limestones, pebble beds, sandstones, etc.). In addition, frost-weathering might have broken some of the rocks up and rendered them more susceptible to mass movements and fluvial erosion.

There is a wide range of evidence that supports the periglacial hypothesis as an explanation for many dry valleys:

1 Many of the dry valleys are fronted by a fan of debris ('head', 'coombe rock'), which is thought to be frost-shattered debris moved by solifluction. In the case of some North Downs dry valleys near Brook in Kent (Kerney et al., 1964), dating of the fans reveals they were formed during the late Devensian Loch Lomond Advance (Zone III). The same applies to the valleys near Wantage, Oxfordshire (Paterson, 1977).
2 Some dry valleys, notably in Wiltshire and Dorset, may contain sarsen blockstreams, the mechanism for the movement of which is probably solifluction.
3 Some dry valleys (see p. 182) have asymmetric slope profiles and cyropediments.
4 The rarity of notches in the chalk scarp crests of southern England may be explained by the fact that most dry valleys are of periglacial origin and thus of no great age. Thus the retreating escarpments have not had time to behead more than a small number of dip-slope valleys (Williams and Robinson, 1983).
5 Many of the dry valleys in the Chilterns do not seem to be related very strongly in their directions to the joints in the Chalk, maybe because they were fashioned by periglacial torrents working on rock that was impregnated with permafrost (Brown, 1969).

Although this evidence for the role of periglacial conditions in fashioning dry valleys appears persuasive, it needs to be remembered that given the great diversity of dry valley forms and locations no single explanation will suffice for all types. In some cases, for example, the role of joint control is clear and spring sapping (which would be of minimal significance under permafrost conditions) appears to have been a major formative influence. The case for the importance of spring sapping has been well put by Small (1965) who points out certain difficulties with the periglacial hypothesis (pp. 3–4):

> In the first place, the most spectacular coombes are fed by surface catchments (the equivalent of snow collecting grounds) which are often extremely limited in area . . . Secondly, the actual courses of many coombes seem inexplicable in terms of the meltwater hypothesis . . . Rake Bottom is incised along a zig zag line down the centre of a broad, gently sloping spur which projects north-westwards from the crest of Butser Hill. This direction is the least likely to be followed by powerful meltwater torrents . . .

The arguments marshalled in favour of the spring-sapping hypothesis include the presence of breaks of slopes along dry valley long profiles (which could be the position of former spring heads); the fact that today some of the most impressive Chalk coombes are occupied in their lower parts by springs; the presence of joint-controlled courses that would be followed as springs cut back; and the preferential development of incised scarp-face coombes at the foot of the Lower Chalk escarpment face in Wiltshire rather than the adjoining Middle and Upper Chalk scarps higher up.

However, the periglacial and spring sapping hypotheses are by no means mutually exclusive. As Small (1965, p. 6) judiciously comments:

> . . . it is not clear whether or not spring action would have been retarded or accelerated under very cold climatic conditions. If at such times the underground circulation of water was able to continue, as a result of periods of summer thaw and infiltration, the damp areas around spring-heads and seepage-points could have been the scene of intense frost-weathering and the process of headward erosion could have been enhanced. Yet again, even if many Chalk valleys were initiated and extended by springs under non-periglacial conditions, considerable modification of the valley sides and heads by periglacial mass-wasting must have occurred during the colder phases of the Quaternary.

Valley meanders and misfit streams

Probably related to just those circumstances that produced dry valleys are another widespread and important landform type – river valleys which appear to be too large to have been moulded by present-day discharges. This was a problem which intrigued W. M. Davis when he visited the Cotswolds (Davis, 1895) and he postulated essentially local mechanisms such as river capture or glacial lake overflow to account for the misfit nature of the valleys of such Thames tributaries as the Coln and the Evenlode (Beckinsale, 1970).

Plate 16. The River Leach, on the Cotswold dip slope, displays evident misfitness to the extent that the present channel is often dry. Note the large size of the ancient valley meanders, and the development of man-made terraces (strip lynchets). (Author)

However, Dury (1954) pointed out the inadequacies of such local mechanisms for such a widespread phenomenon (pp. 81–2):

But meandering valleys are so widely distributed that no system of ice-tongues, ice-fronts, or ice-dammed lakes can be imagined which would account for the observed facts . . . Again, capture cannot explain loss of volume in both

of two opposing stream systems . . . In England valley
meanders are widespread in the Weald . . . beyond the
limits of glaciation – and occur on competing rivers, e.g.
the Warwickshire Avon and its tributaries on the one
hand, and the Cotswold rivers on the other.

In their place he suggested much higher precipitation levels in the
past to account for the high river discharges that moulded the
currently shrunken river networks. He estimated former higher
discharges from the difference in amplitude between the present
river meanders and the larger, ancient valley meanders. Such
calculations are fraught with assumptions and uncertainties and
have been the subject of some debate (e.g. Castleden, 1976;
Beckinsale, 1970; Cheetham, 1980), and not all authors would
concur with Dury (1965) that there had been at least a twenty-fold
decrease in bank-full discharge for rivers in lowland Britain since
Devensian times.

Asymmetric valleys

There are many reports that valley side slopes in parts of Britain
are asymmetric. In some cases, as with the sides of Cheddar Gorge
in the Mendips, the cause may be essentially structural and connected
with uniclinal shifting along dipping strata, but elsewhere there
appears to be a process control associated with periglacial
conditions. Indeed, in many mid-latitude regions of the northern
hemisphere slopes which face west or south-west are usually
steeper than those of other orientations. One explanation for this is
that in such latitudes the greater receipt of insolation and thawing
on south-facing slopes results in greater run-off and down-slope
movement of frost-shattered material. In turn, this is transported
down the valley-side slopes, and the south-facing slopes are
undercut and made steeper by streams pushed towards the south-
facing slopes by the more stable deposits that accumulate on the
opposite slopes (figure 5.8; Catt, 1986a, p. 87).
    Regional valley side-slope asymmetry is well developed in the
Chilterns (Ollier and Thomasson, 1957; French, 1972); in the
North Downs between Canterbury and Folkestone (Smart et al.,
1966); in the South Downs (Williams, 1986); in the North Wiltshire
Uplands, and in the north Dorset Downs (French, 1973, 1976). In

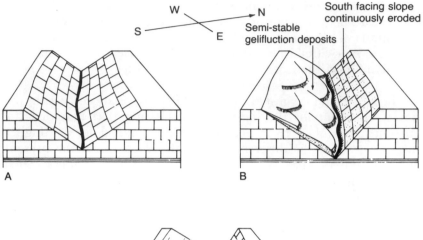

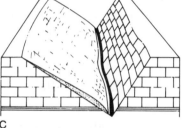

Figure 5.8 The formation of an asymmetric valley (C) from a symmetric valley (A) by unequal periglacial erosion resulting (B) from repeated thawing of the south-facing valley side and removal of its gelifluction deposits by stream action (after Catt, 1986, figure 3.29).

the asymmetric sections of these valleys (generally the middle and lower parts), the gentler slopes (often 7–11°) face north and east, are longer and are often dissected by small, shallow tributary dells. The steeper (often 19–23°) and shorter slopes face south and west and tend to be relatively undissected (Williams, 1980). The asymmetry of slopes is usually accompanied by an asymmetry of periglacial deposits. In the Chilterns, for example, the upper parts of the steep south- and west-facing slopes are usually bare of deposits, whereas the lower slopes and the valley floors are covered with chalky, clayey, solifluction deposits and are full of flints. The gentle north- and east-facing slopes, by contrast, have a tendency to be mantled for their entire length in flinty clay that is largely non-calcareous and derived by solifluction from Clay-with-Flints on the hill crests.

One element of asymmetry is the cryopediment (French, 1973), a concave slope of low gradient which occurs at the foot of the steeper south-west-facing valley-side slopes. These forms are interpreted as replacement slopes of transport which extended outwards as the upper steeper slope retreated, and which are in effect periglacial *glacis* or foot-slopes presumed to develop relatively rapidly on lithologies (like Chalk) that are particularly susceptible to frost-weathering and mass movements.

Another element of asymmetry is the asymmetrical development of smaller tributary valleys. Figure 5.9 shows that in the north Dorset Downs, near Tolpuddle, not only are there steeper west-facing valley-side slopes, but there is a greater density of first order tributaries on east-facing slopes.

It needs to be pointed out, however, that slope asymmetry is by no means restricted to the chalk terrains of southern England. Aspect has been shown to be important in such diverse areas as the gritstone country of north-east Yorkshire (Gregory, 1966b) and the valleys of South Wales where, in the Fforest Fawr, west-facing slopes are markedly steeper (Crampton and Taylor, 1967). There is even evidence for asymmetry in fluvial valleys that formed at times of low sea level on the floor of what is now the North Sea (Long and Stoker, 1986).

River terrace aggradation

River terraces are developed along many major British rivers, including the Thames (Gibbard, 1985); yet phases of river channel aggradation, leading to terrace formation, are notoriously difficult to explain in any simple way (Goudie, 1983, p. 21):

> They may sometimes form as a result of non-climatic causes, such as tectonic change, sea-level change, glacial invasion of catchment areas and so forth. However, even if one can eliminate non-climatic causes, it is difficult to draw precise inferences as to climate from alluvial stratigraphy within terraces because of the variety of possible climatic influences: amount of precipitation, distribution of precipitation throughout the year, mean and seasonal temperature, and other climatic variables. Moreover, the response of a stream, in terms of load and

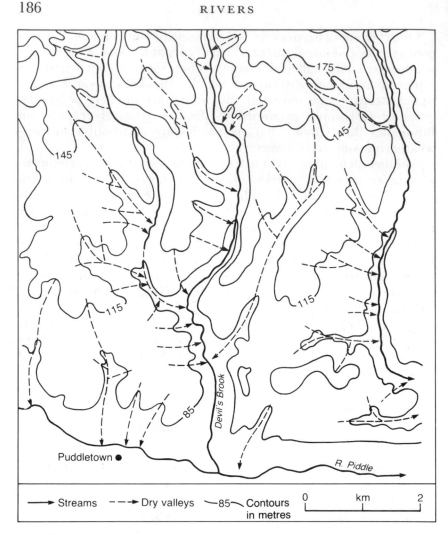

Figure 5.9 Asymmetrical terrain in the chalk of the north Dorset Downs near Tolpuddle. Note the steeper west-facing valley side slopes and the asymmetric tributary patterns (after French, 1976, figure 13.2).

discharge, to changes in such climatic variables, will be influenced by vegetation cover, slope angle, the range of the altitude of the basin, and other circumstances.

One of the traditional simple explanations of suites of terraces was to relate them to Quaternary sea-level changes, with drainage incision taking place at times of low sea-level during glacials, and drainage aggradation taking place at times of high sea-level during interglacials.

This thalassostatic interpretation has some intuitive appeal, not least because many studies using such techniques as boring and seismic investigations have revealed buried channels in the lower stretches of many rivers and palaeodrainage channels on the floor of the North Sea basin and other parts of the continental shelf (see p. 284).

In recent years, however, it has become apparent that in the middle and upper reaches of many rivers phases of aggradation were related to periglacial conditions. This type of interpretation (see figure 5.10) has been applied by Briggs and Gilbertson (1980) for the Upper Thames valley. They find that the terrace sediments there contain remains of cold condition mollusca and pollen, include fossil ice-wedge casts and involutions indicative of permafrost (Seddon and Holyoak, 1985), and contain gravels which are for the most part subrounded to subangular in shape and, given the soft readily abradable nature of the Jurassic limestone of which they are composed, appear to have been transported only a limited distance before deposition. The sediments also show numerous small

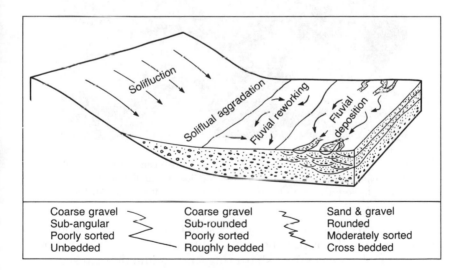

| Coarse gravel
Sub-angular
Poorly sorted
Unbedded | Coarse gravel
Sub-rounded
Poorly sorted
Roughly bedded | Sand & gravel
Rounded
Moderately sorted
Cross bedded |

Figure 5.10 Diagrammatic representation of processes causing Pleistocene terrace aggradation in the upper Thames basin (after Briggs and Gilbertson, 1980, figure 2).

Plate 17. River terrace materials of late Pleistocene age (probably mid to late Devensian age) near Oxford, close to the floodplain of the Thames. Note the cross bedding and the ice wedge cast. (Author)

channel infills and other structures which indicate that deposition occurred under braided conditions. Thus the model of deposition is one in which intense frost-shattering and solifluction caused river systems to become choked with locally derived debris, which was then reworked by dominantly braided-stream systems, or, in some cases, by gravel-bed meandering rivers (Bryant, 1983).

Such phases of periglacial aggradation may have recurred repeatedly. In their study of the palaeobraided river deposits of the Blackwater valley and its tributaries in Berkshire and Hampshire, for example, Clarke and Dixon (1981) claimed to find 11 terraces spanning the last 900,000 years, each of which, they believed was the result of periglacial aggradation.

The general significance of periglacial fluvial activity has been summarized by Worsley (1977, p. 213):

> Fluvial activity is a vital component of contemporary periglacial environments . . . Within the former British periglacial zones there can be little doubt that the dominant element of the landscape was the fluvial erosional-depositional system. Many stratigraphic studies of fossiliferous river terrace sequences, in conjunction with radiocarbon dating, have revealed the considerable extent of fluvial aggradations during the Devensian, especially beyond the direct effects of glaciation per se. Sedimentological studies show that for most of the last glacial stage, run-off was reworking the head on valley slopes and transporting it on to wide alluvial tracts across which rivers migrated to and fro.

However, in addition to the importance of periglaciation, glaciation as a cause of river aggradation also has to be considered. This applies, for example, to the River Severn (Dawson and Gardiner, 1987), a river whose catchment was much influenced by glacial events of the Pleistocene. The oldest deposits, the Woolridge Gravels of Wills (1938), were regarded by Hey (1958) as the product of Wolstonian glacial outwash. The Main Terrace, which can be traced from Bridgnorth to Gloucester, contains a sizeable proportion of erratic material derived from 'Irish Sea' glaciation and may be interbedded with flow-till. Its aggradation may have developed, at least partially, as outwash from the late-Devensian ice sheet at its maximum position (Shotton, 1977). The lower Worcester Terrace also contains a significant amount of large clasts of Irish Sea erratic material and may be related to a later re-advance of the Devensian ice sheet.

It is not, however, possible to interpret British river terraces solely in terms of cold climate conditions, be they periglacial or glacial. Some terraces (e.g. the Beeston Terrace of the Trent) are of interglacial age, and the mechanism of aggradation by braided rivers is not the only one that may be inferred. Indeed Clayton (1977) has hypothesized that some British river terraces are the dissected cut and fill floodplains of meandering channels not unlike those forming floodplains today. Terraces remain one of the least understood fluvial features in Britain.

6

SLOPES AND MASS MOVEMENTS

Slope forms

For much of the last 150 years, British geomorphologists have been more concerned to study essentially flat areas as part of their pursuit of erosion surfaces (see pp. 1–8), than they have to find out the true nature of slope forms and trends of slope evolution. Until the 1960s most discussions of slope form were essentially descriptive, subjective and non-quantitative.

For example, the form of chalk-slopes has frequently been expressed in qualitative terms ('ever-recurring double curves', 'lines of beauty', 'whale-back downs', etc.) which emphasize their rounded nature. Quantitative data are generally sparse, although a notable exception is provided by the work of Clark (1965) who, on the basis of his detailed measurements proposes that, 'the myth of the "characteristic simple convexo-concave Chalk slope" can be confidently discarded' (p. 32). Only about 32 per cent of his measured profiles had the simple convexo-concave form so often proposed as the 'normal' profile for chalk-slopes. Rectilinear segments were common. Few chalk-slopes have angles greater than about 35°, and slopes greater than 32–33° show signs of instability, such as fractures in the turf cover and well-marked terracettes. A peak of around 29–30° seems to be the maximum repose angle for chalk-slopes in southern England.

A rather comparable attack on orthodox views of slope form, which had been rooted in the tradition of Davisian slope decline and peneplanation, came from Dury (1972) who argued that certain slope forms in southern England are essentially pediments.

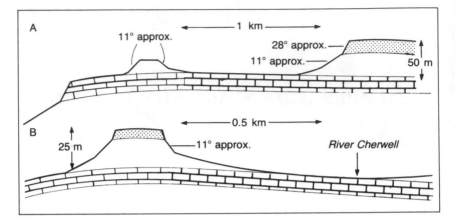

Figure 6.1 Erosional slopes on escarpment and outliers developed on Jurassic strata in Northamptonshire (after Dury, 1986, figure 62).

He defines a pediment as – 'a degradational slope, cut across rock in place, abutting on a constant slope at its upper end, and decreasing in gradient in an orderly fashion in the downslope direction' (p. 14). However, while he provides measured profiles which indicate that if stripped of their soil and vegetation certain hills in southern England would not 'look at all out of place on a desert pedeplain' (p. 147) he recognizes certain differences between his pediment slopes and those that are held to be characteristic of desert areas. Among these are a continuous soil cover and the absence of a pediment angle – the sharp break between pediment-head and hill-slope. None the less, as figure 6.1 shows in the case of the slopes of the Jurassic escarpment and its outliers in Northamptonshire, a model of slope evolution involving parallel retreat is more appropriate than a Davisian one which involves slope decline through time. The same applies further south-west along the Cotswold escarpment in Gloucestershire, where a large number of outliers (inselbergs?) show similar slope forms to the escarpment, from which they have long since been separated.

Young (1961) attempted to see whether in England and Wales the nature of slope retreat is such that certain characteristic angles are formed more frequently or remain unaltered for longer periods, than others. To this end he measured valley-side profiles in North Devon, Central Wales and the Southern Pennines (figure 6.2). He found a tendency for a group of characteristic angles at 25–29° and suggested that, following stream rejuvenation in an area and

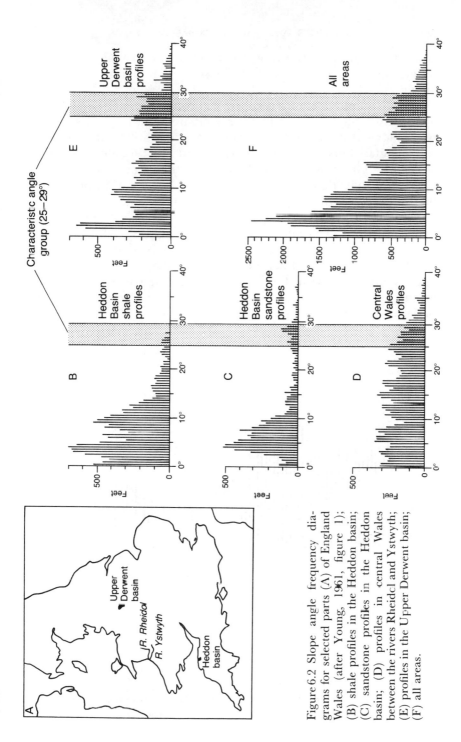

Figure 6.2 Slope angle frequency diagrams for selected parts (A) of England Wales (after Young, 1961, figure 1); (B) shale profiles in the Heddon basin; (C) sandstone profiles in the Heddon basin; (D) profiles in central Wales between the rivers Rheidol and Ystwyth; (E) profiles in the Upper Derwent basin; (F) all areas.

associated basal undercutting, these are the first category of slopes to develop that are not rapidly reduced by denudation to a gentler angle.

While Young (1961) used surveyed profiles, Gregory and Brown (1966) attempted to identify characteristic angles by constructing morphological maps and then measuring the areal extent of each slope angle. This work was carried out in Eskdale, north-east Yorkshire. The most striking features of their findings (see figure 6.3) were that there is an absence of large variations in the distribution of angular slope angles and that, as might be anticipated, there is a decrease of area as the angular value increases, with the maximum value occurring at 1°, while at values greater than 23½° some angles are not represented at all. They also found that there were differences in slope angles on different rock types (figure 6.4) with, for example, resistant Dogger sandstones forming quite steep free faces and the Kellaways outcrop supporting the least striking slopes in the area.

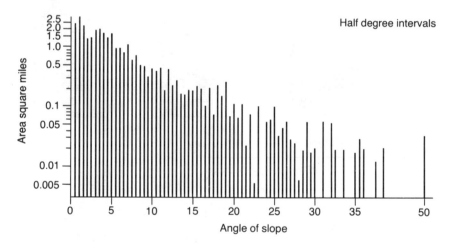

Figure 6.3 Angle–area relationships for sample drainage basins in the Esk valley, north east Yorkshire, based on ½° slope intervals (after Gregory and Brown, 1966, figure 5A).

Tinkler (1966) measured slope profiles on the Eglwyseg escarpment in North Wales, which are developed on Carboniferous limestone. He found free faces developed in resistant limestone beds with angles over 50°, and characteristic angles for bedrock slopes of

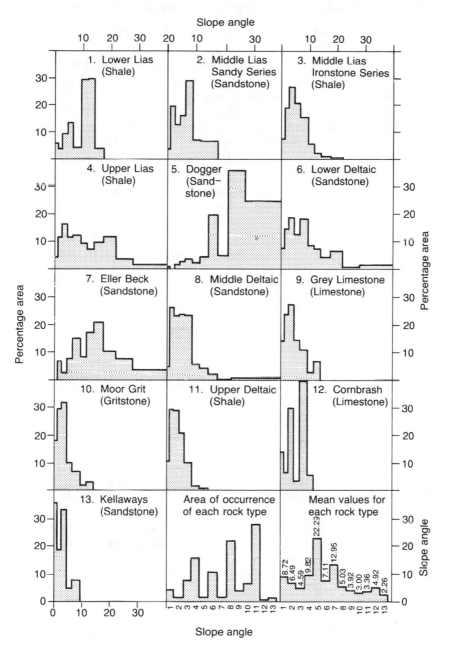

Figure 6.4 The angle–area relationships of each rock type in the sample basins of the Eskdale basin, north-east Yorkshire (after Gregory and Brown, 1966, figure 8).

37–39°. Screes had characteristic angles of 35° and colluvial clitter slopes angles of 29–31°.

The next major study to seek characteristic threshold slopes was that of Carson and Petley (1970), using field survey of slope angles on Exmoor (middle Devonian strata) and the south Pennines (Shale Grits). The main finding was that the frequency distribution of angles of straight hill-slopes was trimodal in form and angles of 20–21°, 25–27° and 32–34° were characteristic of both areas. In addition, however, Carson and Petley analysed the shear strength characteristics of the waste mantles on their slopes, and on this basis suggested that these particular characteristic angles are probably limiting angles of stability for three types of waste mantle. Partial confirmation of this was provided by Rouse and Farhan's analyses (1976) of slope angles and material properties on valley side-slopes in South Wales, where actual and predicted mean angles were around 17°.

Slope evolution

Building upon such measurements of slope form, Carson and Kirkby (1972, pp. 319 ff) provide a three-stage model of slope profile development for humid temperate areas such as the British Isles. In the first stage stream incision occurs into a land surface, producing straight hillside slopes inclined at a particular angle of stability related to the properties of the material making up the slope. In the case of competent, resistant beds vertical cliff faces might develop in response to rapid incision. With a slower rate of downcutting, weathering of the rock surface might result in the production of a talus slope, maintained at the angle of repose of the talus. In areas where slopes were made of less competent material, even if they maintained particular angles of straight slope during stream incision, slopes would be gentler.

In stage two, when downcutting by streams ceases to be rapid, slopes begin to develop independently of stream action. The pattern of development depends upon the relative importance of three groups of processes: weathering and mass failure; surface and subsurface soil wash and solution; and soil creep. The first straight slope produced by stream incision becomes modified. The mantle on the hillside slope undergoes progressive changes due to weathering, which affects the stability of the slope by altering such qualities as cohesion, permeability and internal friction. Different threshold slopes will develop through time. For example, in strong, well-jointed rocks, the first stable, debris-mantled slope will be a talus slope at the angle of respose of the debris (*c*.32–38°). Subsequently, continuing weathering causes this mantle to become unstable and the slope angle declines to around 25–28°.

In stage three, once the soil mantle becomes so weathered that little further change takes place and the slope adjusts to the angle of stability of this type of soil, the other groups of processes begin to emerge as important: wash, creep and solution. These cause the development of a convex profile at the summit of the slope and of a basal concavity. As development proceeds, these two elements encroach on the stable straight slope and eventually convert the entire profile to a convexo-concave form. The convexo-concave profile gradually becomes reduced in angle, and the rate of development very slow. The ultimate result may be the production of a landscape dominated by low angle slopes – in effect, a Davisian peneplain.

However, it is extremely difficult to test such a model of slope evolution since it is virtually impossible to observe the effect on hillslopes of extended periods. One method that tries to overcome this problem is ergodic reasoning – the substitution of space for time. The classic example of such an approach is Savigear's (1952) study of slope evolution in South Wales.

In his study area, Savigear had high cliffs (120–150 m) developed in Old Red Sandstone. Some of them were being actively attacked and undercut by marine erosion, but others had become shielded from continued marine attack by the growth of spits. The protecting spit had developed progressively from west to east, so that those at the western end had been sheltered for a long period and those at the eastern end for a much more limited period. A comparison of the most recently sheltered slopes with the oldest slopes in the sequence suggests that, over time, a relatively steep straight slope is replaced by another at a lower angle (figure 6.5).

Landslides: geological controls

Although under present climatic conditions slope instability and major mass movements may be less severe and extensive than they were, for example, in the late Pleistocene (see p. 96), slopes in

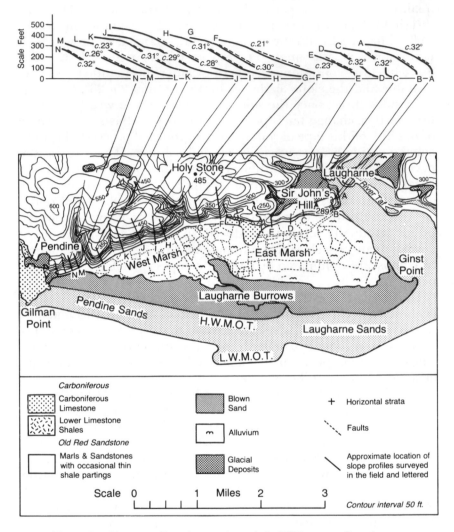

Figure 6.5 Slope profiles along a degraded cliff line near Laugharne, South Wales (modified from Savigear, 1952).

England and Wales are frequently far from stable. Landsliding is of widespread occurrence, though there are clear lithological and structural controls of distribution and form of failure. Certain formations appear especially prone to slides (see, for example, Jones et al., 1988).

TERTIARY BEDS

Mass movement features are plentiful in the Tertiary sediments of south-east England. The London Clay slopes on the southern side of the Thames Estuary have failed in a spectacular manner on the Isle of Sheppey and at Herne Bay (Bromhead, 1978) and also in Essex, notably at Hadleigh (Hutchinson and Gostelow, 1976). Landsliding also affects the Palaeogene rocks of the Hampshire Basin (e.g. at Bouldnor on the north-west coast of the Isle of Wight) and at Barton in Hampshire where Barton Sands (and Plateau Gravels) overlie Barton Clays.

THE LOWER CRETACEOUS

The *Upper Greensand* of the Lower Cretaceous frequently forms landslips where it overlies the Gault Clay. Some spectacular coastal landslide complexes have developed in this situation of which notable examples are those of the Dorset Coast at Fairy Dell, Stonebarrow Hill. Inland slopes along the Char Valley are also affected (Brunsden and Jones, 1976). The massive undercliff of the southern coast of the Isle of Wight is a great slipped area which also involves the Gault and the Upper Greensand (Hutchinson, 1987).

The *Lower Greensand* scarp of the Weald also appears to have been highly susceptible to landslipping. The Hythe Beds scarp, which consists of a thin capping of chert, called Rag-and-Hassock, resting on Atherfield and Weald Clays, appears to have suffered repeatedly from slumping along much of its length, with failure of the upper slopes resulting in debris moving downslope over the Clays to form a chaotic, irregular, hummocky footslope of slumped blocks and lobes (Jones, 1981). Famous rotational landslips that have occurred are those at Bower Hill and Tilburstow Hill in Surrey (figure 6.6; Gossling 1935; Gossling and Bull, 1948).

Within the High Weald, the sandstones of the *Hastings Beds* (especially the Tunbridge Wells Sands) have frequently slipped over the underlying Wadhurst Clay, and the distribution of such mass movement phenomena is clearly depicted in figure 6.7. Landslipping takes place today on slopes of more than 12°, but under periglacial conditions much lower angle slopes were affected (Robinson and Williams, 1984).

INFERIOR OOLITE

The most widespread incidence of landsliding in Britain probably occurs on the Jurassic Inferior Oolite scarps. These limestones and

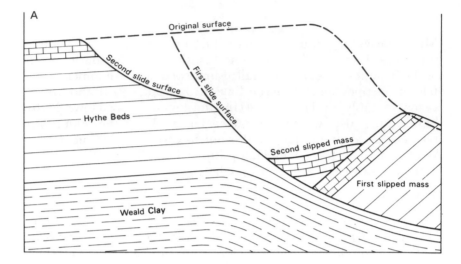

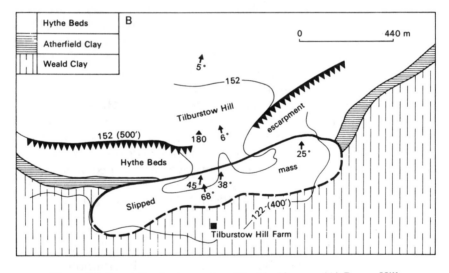

Figure 6.6 Landslips in the western Weald of Surrey: (A) Bower Hill;
(B) Tilburstow Hill (after Sparks, 1971, figures 4.11 and 4.12).

sandstones overlie Lias Clays. The Cotswold escarpment is affected
for much of its length from the Bath area northwards (Chandler et
al., 1976); other important slips occur through Northamptonshire
and Rutland (Chandler, 1976) and into Lincolnshire (Penn et al.,
1983).

However, within the Cotswolds there are clear patterns of landslide distribution, related to lateral changes in geology along the escarpment (Hawkins and Privett, 1979). For example, south of the Frome Gap, as far south as Dyrham, landsliding is relatively infrequent. The reason for this, according to Butler (1983), is that in that stretch of the escarpment the Upper Lias Clay is thin and poorly developed and its place is taken by the much more stable Cotteswold Sands. The reappearance of large-scale landsliding south of Dyrham is due to the replacement of the Cotteswold Sands by the siltier, more incompetent and less ferruginous Midford Sands, and to the fact that incision has occurred deep into the Lower Lias Clay because of the thinness of the formations beneath the Fuller's Earth Beds. North of Stroud, the Upper Lias Clay is the dominant bed beneath the Inferior Oolite, and this highly argillaceous, soft clay is the cause of the almost ubiquitous instability in the north Cotswolds. Another bed which promotes instability is the Fuller's Earth Clay and such failures as there are between Stroud and Bath often develop in this stratum.

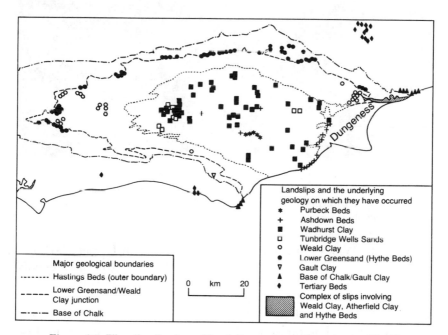

Figure 6.7 The distribution of landslips and mudflows in the Weald and surrounding areas of south-east England (after Robinson and Williams, 1984, figure 8).

UPPER CARBONIFEROUS GRITS AND SANDSTONES

In the Pennines, sandstones and grits have frequently slipped over underlying mudrock formations. For example, the Mam Tor and Edale End landslides (Redda et al., 1985) occur where Mam Tor Beds (thin sandstones and shales) overlie Edale Shales. The same applies at Alport Castles (Johnson and Vaughan, 1983), one of the largest landslides in inland Britain, where massive sandstones (Shale Grits) overlie less competent mudstones and thin sandstones (Mam Tor Beds) and Edale Shales.

Plate 18. Alport Castles in the Pennines, produced by the failure of massive sandstones over less competent mudstones and thin sandstones. The main escarpment is to the right, and the Alport Valley to the left. (Author)

In South Wales landslipping also takes place in Upper Carboniferous rocks. Within the coalfield landslipping is not evenly distributed, being most frequent in the north-east of the area (e.g. in the Ebbw, Sirhowy and Rhymney valleys) and in some of the valleys of the central coalfield (e.g. Rhondda, Ogwr, Garw and Llyfni valleys). The majority of the landslipping occurs at or close to the junction of the Upper Coal Measures (Pennant Sandstone) with the Middle Coal Measures. The latter is an essentially argillaceous sequence, often associated with springlines (Forster and Northmore, 1985).

The Etruria marl (Upper Coal Measures) of the Midlands is another argillaceous material in which landsliding occurs (Hutchinson et al., 1973).

MISCELLANEOUS

Although attention has been drawn here to certain formations which are especially characterized by landslip activity, landslips are recorded from other strata including Permo-Triassic conglomerates and mudstones in Devon (Grainger and Kalaugher, 1987), Upper Carboniferous mudrocks of the Upper Culm, also in Devon (Grainger and Harris, 1986), and Devonian sandstones and mudstones in Gwent (Early and Jordan, 1985). At the other end of the age spectrum, Palaeocene deposits, including tills, have also been affected. Warren et al. (1984), for instance, have drawn attention to the way in which many drumlins in North Wales are much modified by landslipping, while Conway (1979) has highlighted the way in which water-bearing head deposits have promoted landsliding in West Dorset. In South Wales, most of the landslides in the Coalfield are shallow failures, confined to the glacial head deposits, and translational in nature (Rouse and Bridges, 1985).

The nature of landslides

Many landslides are associated with the presence of mudstone outcrops: Keuper Marl, and the Lias, Oxford, Kimmeridge, Weald, Gault and London Clays. These are stiff, fissured, heavily overconsolidated materials which are characterized by time-dependent loss of strength. Such overconsolidated clays possess low compressibility, moisture content, permeability and sensitivity as a result of burial and loading beneath younger deposits. Erosion of the overburden of such younger deposits has sometimes been considerable (up to 1,000 m in the case of the Lias and 400 m in the case of the London Clay), meaning that mudrocks exposed at the surface experience normal stresses much lower today than in the geological past (Anderson and Richards, 1981; Cripps and Taylor, 1986, 1987).

Exposure at the surface results in a progressive loss of strength from the peak strengths of the overconsolidated clay to a residual strength which may be weaker than that of a normally consolidated clay. This loss of strength reflects several influences which result from readjustment to the stress environment on exposure. Weathering processes create mineralogical and chemical changes which lead to a loss of cohesion. In addition, stress relaxation as the overburden erosion continues causes rebound (dilation) joints and fissures to open up parallel to the developing topography. This facilitates the ingress of water into the rock with the result that expandable clay minerals swell and slake, thereby encouraging the general softening and swelling of the mudrock as a whole.

The precise form of landslide which occurs on any particular clay slope seems to be closely related to slope angle (Chandler, 1970). As figure 6.8 shows, on both London Clay and Upper Lias slopes rotational slides tend to be prevalent on steeper slopes (normally greater than about 13°), whereas successive rotational slides, non-circular slides, slab slides and undulations occur on progressively lower-angle slopes. The limiting angle for these types of movements on Upper Lias Clays is 9° and the maximum stable slope is about 10½°. The corresponding angles for the London Clay are 8° and 10°.

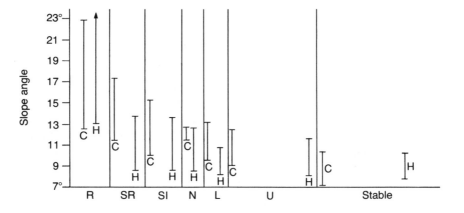

Figure 6.8 Critical angles for the development of various types of mass movements on mudrocks in England (after the work of Hutchinson (H) and Chandler (C): R = rotational slides; SR = successive rotational slides (ridges); SI = successive rotational slides (hummocks); N = non circular slides; L = slab slides; U = undulations.

Multiple rotational slips tend to occur preferentially where a permeable but coherent material overlies a less permeable thick clay. The importance of such a caprock is that it probably retards degradation of the scar of the first slip. As Brunsden (1979, p. 168) explains:

> Thus the stresses on that slope are only slowly diminished with time, unless erosion at the base of the slope is able to destroy the original slide and steepen the scarp face. The cap-rock will slow down this process and maintain a steep under slope so that, as the original slip slowly moves downslope, or is degraded, the stresses for failure can re-occur. It is therefore common to find that there are long intervals between the individual failures. Sufficient time has to elapse before the original failure has moved or degraded to recreate an unstable situation.
>
> In British most of the multiple landslides appear to be inactive, or to show movements of old blocks only. This is probably explained by the fact that the transport and erosion processes at the toe are now very slow, whereas in the late-glacial period when many of these slides are thought to have commenced, erosion rates and pore pressures would have been higher.

Translational slides are another widely occurring form of mass movement. Such types are very long compared to their thickness and surmount what is an essentially planar, slope-parallel glide surface. They can travel far, and the ratio of thickness to downslope length tends to be between 0.03 and 0.06. By contrast, the thickness length ratios of rotational slides are between 0.15 and 0.33. An excellent example is provided from the Cotswold outlier of Bredon Hill (figure 6.9).

Mudslides, which are relatively slow-moving lobes or enlongate masses of softened, argillaceous debris that advance chiefly by sliding on discrete boundary shear surfaces, occur in overconsolidated clays on low-angle slopes (usually 5–15° but sometimes as low as 3–4°). They may occur as groups as on the west Dorset coast near Lyme Regis. They tend to occur as three main units: an upper supply area, a central neck or track, and a lobate toe. Movement rates are generally slow and seasonal, with increasing rates of movement occurring after high rainfall events. Rapid surges may occur, possibly when loading of the head is caused by material descending from feeder slopes above.

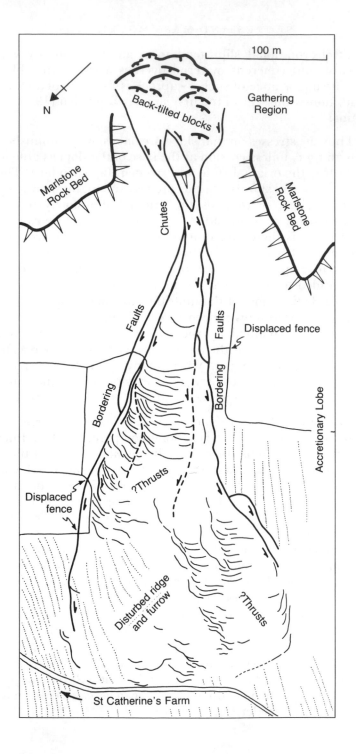

Landslip history

One might intuitively expect land-sliding activity in England and Wales to have taken place with greater relative frequency at certain phases in the past:

1 during times of glacial oversteepening of valleys;
2 during times of drainage incision (to produce what are now dry or misfit valleys);
3 under periglacial conditions when cambering and valley bulging occurred (see p. 96);
4 at times when permafrost was waning (e.g. between 11,000 and 9,000 years BP);
5 when climate was more humid (e.g. in the Atlantic phase of the Holocene);
6 during cold phases of the Holocene (e.g. the Little Ice Age).

Ideas of this sort have received support from time to time (e.g. Starkel, 1966), but reliable dating evidence is still scarce. Johnson (1987, p. 594) suggests, on the basis of such reliable data as there are, 'that major rockslides have not occurred with any marked periodicity' and that high rainfall events which trigger such mass movements need not have been associated with secular wet periods.

None the less, some landslips have a long and spasmodic history. The best available study of a single landslip's history is probably that of Hutchinson and Gostelow (1976), which was carried out at Hadleigh in Essex. Hadleigh Cliff is developed in London Clay and was originally formed by strong fluvial erosion in the Middle and late Devensian. Such erosion had virtually ceased by the latter part of the late Devensian (around 15,000 BP) and since then the cliff has degraded episodically; Hutchinson and Costelow see this episodicity as a response to climatic changes. By dating extensive layers of colluvium beneath the degraded cliff they were able to recognize four main stages of degradation, with intermediate periods of relative slope stability:

1 late-glacial, periglacial mudsliding (completed by c.10,000 BP);
2 early Atlantic mudsliding (c.7,000–6,500 BP);
3 early sub-Atlantic mudsliding (c.2,100–2,000 BP);

Figure 6.9 A translational slide on Bredon Hill (after Allen, 1985, figures 9.15 and 9.16).

4 recent landslipping (notably in the late nineteenth century), possibly caused in part by human interference.

Some other forms of mass movement

ROCK FALLS

Rock falls involve the free movement of material away from a steep slope. For that reason they are essentially restricted in their distribution to steep slopes, including free faces and marine cliffs. Rock falls undoubtedly occur in England and Wales and are a major source of material for scree formation, but there are surprisingly few records of their occurrence.

The causes of the failure that produces rock falls include undercutting of slopes by erosion. However, the most common causes of small falls in Britain are probably joint enlargement and block loosening caused by freeze–thaw activity. Confirmation of this comes from some work done on chalk falls on the Kent coast by Hutchinson (1971) who shows that falls between 1810 and 1970 were closely related in their occurrence and frequency to average number of days of air-frost per month and to the average monthly effective rainfall (figure 6.10).

SNOW AVALANCHES

Snow avalanches are relatively rare today in England and Wales, although they may have been more frequent in the Pleistocene, possibly contributing to chute development on hillsides (see p. 95). However, to judge from Ward's (1984) work in Scotland, even today they may be rather more widespread, frequent and large than often maintained, and in areas like the Cheviots may create dangers for climbers and hill walkers. For example, on 7 February 1988, two people were killed and three others injured on Bizzle Crags in Northumberland; another was injured after an avalanche swept him down Sca Fell in the Lake District. In human terms probably the most serious snow avalanche in England was that which hit Lewes in Sussex on Christmas Eve, 1836, sweeping away a number of cottages and killing eight inhabitants. The site of the calamity is marked by the present 'Snowdrop Inn'; the snow fell from the top of the South Downs behind (*Sussex County Magazine* 1, 1927, 70–4).

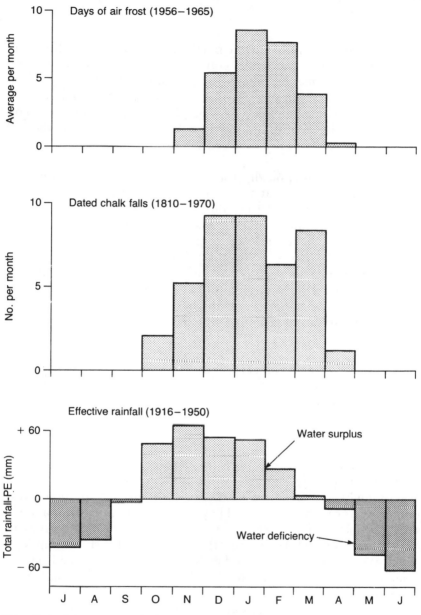

PE = Potential evapo-transpiration

Figure 6.10 The relationship between rock falls, days of air frost and effective rainfall for chalk cliffs on the Kent coast (after Hutchinson in Brunsden, 1979, figure 51.14).

BOG BURSTS

Mires may be susceptible to one particular type of slope failure – the bog burst. As discussed in the context of peat erosion (see p. 300) many peat bogs may become potentially unstable as they build up their levels so that if a severe stress is applied, as for example by a phase of extreme rainfall, they might fail spontaneously and catastrophically. A graphic account of a failure which affected Solway Moss is provided by Lyell (1835, p. 208):

> This same moss, on the 16th of December, 1772, having been filled with water during heavy rains, rose to an unusual height, and then burst. A stream of black half-consolidated mud began at first to creep over the plain, resembling, in the rate of its progress, an ordinary lava current. No lives were lost, but the deluge totally overwhelmed some cottages, and covered 400 acres. The highest parts of the original moss subsided to the depth of about twenty-five feet; and the height of the moss, on the lowest parts of the country which it invaded, was at least fifteen feet.

Peat mosses are also subject to sliding activity, particularly after intense summer storms. Notable examples have been reported from the Pennines (Crisp et al., 1964; Carling, 1986).

DEBRIS FLOWS

Debris flows involve the mass movement of a wet mixture of granular solids, clay minerals, water and air under the influence of gravity, where intergranular shear is distributed more or less uniformly through the mass. They occur predominantly on steep slopes in upland areas, including the Old Red Sandstone escarpment of southern central Wales (Statham, 1976), the scree slopes of North Wales, the Pennines, and at Borrowdale in the Lake District. They can be recognized in the field by the presence of small, bare patches of ground at their heads; their steep, shallow courses; their raised edges composed of coarse levée material; and overspilling debris lobes and fans towards their bases.

There are various records of such flows being triggered by high-intensity summer storms (e.g. Addison, 1987; Carling, 1987; Wells and Harvey, 1987).

TERRACETTES

A common feature on many British hill-slopes is the presence of suites of roughly parallel small terraces or ridges which extend across the slope more or less normal to the direction of maximum slope. They are termed terracettes. Their origins are still controversial (Vincent and Clarke, 1976). Some basic morphometric data averaged for 215 sites in western Britain are provided by Anderson (1972):

Tread angle	10.5°
Riser angle	39.9°
Tread width	54.5 cm
Riser width	90.5 cm

In some cases they may be animal tracks or they may be accentuated by animal movement along them once they are formed in some other way. However, some examples occur in areas where animals are very rare, suggesting that some other mechanism needs to be inferred, such as soil creep, solifluction or minor rotational slipping. One also needs to explain why terracettes occur in some places and not in others.

In a study of terracettes on different lithologies in Exmoor and the southern Pennines Carson (Carson and Kirkby, 1972, pp. 174–6) found that their occurrence was closely linked to the thickness of the soil mantle on the hill-slopes. They occur preferentially where the slope mantle is very shallow. Where the mantle is thicker other indicators of instability may develop such as scars and slides. This may be because with thicker mantles few binding plant roots extend into bedrock, with the result that there is a layer between the vegetation mat and the underlying rock surface in which the shear strength is at a minimum. He thus suggests that 'terracettes occur where the shear strength of the waste mantle alone is insufficient to maintain the slope at its particular angle, and that pseudo-stability arises from the extra strength implanted by the vegetation mat' (p. 176).

Soil creep

Soil creep, although an essentially unspectacular process, may be an important form of denudation on slopes that are too gentle for the more dramatic forms of mass failure and show no clear evidence of erosion by running water. In the sod-covered landscape of much of England and Wales the continual and long-continued operation of freeze and thaw, growth and decay of plant roots, and the action of earthworms, ants and various animals may cause creep to move appreciable amounts of soil material down hillsides. It has often been held responsible for development of low-angle summit convexities and for the formation of terracettes.

The evidence for soil creep includes both environmental indicators and field measurements. The former category of evidence includes outcrop curvature, tree curvature, turf rolls downslope from creeping boulders, displaced posts, poles and monuments, and stone lines at the approximate base of creeping soil. However, as Finlayson (1985) has cautioned, such indicators may be explained in other ways.

Some of the empirical results obtained by field measurements are listed in table 6.1. Compared with those from some other parts of the world the rates, expressed volumetrically in $cm^3 \, cm^{-1} \, y^{-1}$, are not especially high, clustering around 0–5.

A detailed study has been made of the controls of rates of soil creep in upper Weardale, northern England by Anderson and Cox (1984). This is a grassland area, underlain by Carboniferous sedimentaries. Six different methods of soil creep rate determination were made on 20 plots, and some 20 soil properties were also determined. Their major finding was that the moisture-related soil variables (e.g. field capacity, winter and summer moisture levels, shrinkage and plasticity) are the most important controls of creep rates in their catchments. However, another major discovery was that there was only a very limited relationship between rates of soil creep and slope gradient as represented by the size of slope angle. Moisture appeared to swamp any influence that slope might have. This confirmed the work of Finlayson (1981) who also reported a lack of relationship between rates of soil movement and slope angle on the East Twin catchment in the Mendip Hills.

Rates of soil creep were measured in the Upper Wye valley of mid-Wales by Slaymaker (1972), using grassland sites. He also found (p. 54) 'that the effect of slope is so small in comparison with that of the dynamic soil characteristics that it can be virtually ignored in analysing variations in rates of soil creep'. His rates varied from 0 to 11.8 $cm^3 \, cm^{-1} \, y^{-1}$. However, he did find that those slopes being actively undercut by stream erosion at their base had high rates (on average 3.86 $cm^3 \, cm^{-1} \, y^{-1}$) compared with

Table 6.1 Rates of seasonal soil creep

Location	Source	Rate $(cm^3 cm^{-1} y^{-1})$
England and Wales		
Central England	Chandler and Poole (1971)	0.3
Mendips	Finlayson (1976)	1.12–5.09
Wales	Slaymaker (1972)	2.7
Derbyshire	Young (1960, 1978)	0.5–0.61
Weardale	Anderson and Cox (1984)	0.75
Other Countries		
Maryland, USA	Various, in table 5.5 of Brunsden (1979)	1.3
Ohio, USA	Various, in table 5.5 of Brunsden (1979)	6.0
Kuala Lumpur, Malaya	Various, in table 5.5 of Brunsden (1979)	12.4
Scotland	Various, in table 5.5 of Brunsden (1979)	2.1
Puerto Rico	Various, in table 5.5 of Brunsden (1979)	1.53
New Mexico	Various, in table 5.5 of Brunsden (1979)	4.9
New Zealand	Various, in table 5.5 of Brunsden (1979)	3.2
South Alaska	Various, in table 1 of Young (1974)	15.0
Washington DC, USA	Various, in table 1 of Young (1974)	0.2
Tatar, USSR	Various, in table 1 of Young (1974)	5.7–8.4
New South Wales, Australia	Various, in table 1 of Young (1974)	1.93–3.2
Northern Territory, Australia	Various, in table 1 of Young (1974)	4.4–7.3

slopes which were protected from the action of basal streams by colluvial and alluvial deposits (on average 1.06 cm^3 cm^{-1} y^{-1}).

The role which organisms have played in mobilizing debris on slopes in England and Wales is a topic which has probably been grossly under-researched, although important preliminary investigations on worms were carried out over a century ago by Charles

Darwin (Darwin, 1882). He reported on the way in which marker layers (e.g. coal dust or chalk) were gradually covered up; large stones gradually sink downwards; substantial quantities of soil were moved for cast formation; mineral grains were ground fine in their gizzards; and casts were degraded by wind and water. Even under the sod-covered landscape of England and Wales their activities go on apace. As Darwin (1882, p. 316) concluded:

> When we behold a wide, turf-covered expanse, we should remember that its smoothness, on which so much of its beauty depends is mainly due to all inequalities having been slowly levelled by worms. It is a marvellous reflection that the whole of the superficial mould over any such expanse has passed and will again pass, every few years through the bodies of worms. The plough is one of the most ancient and most valuable of man's inventions, but long before he existed the land was in fact regularly ploughed, and still continues to be thus ploughed by earth-worms.

7

KARST

Introduction

Limestones of great diversity outcrop over large areas of England and Wales (Sweeting, 1972; figure 7.1). The largest areal extent of limestones in Britain is composed of Mesozoic rocks, including the dolomites of the Permian, the Oolites of the Jurassic, and the Chalk of the Cretaceous. However, the British limestones have a tendency to be less porous and less permeable with age, so that karst landforms are best developed (though not to the exclusion of younger rocks) on the extensive Carboniferous limestones and on Devonian beds. Indeed, the most important karst areas in Britain occur on the Carboniferous limestone in north-west Yorkshire, Derbyshire, the Morecambe Bay area, the Mendips, the South Wales coalfield, and North Wales. The Devonian karst area in south Devon, is of modest extent, but has some important caves and solution pipes (Brunsden et al., 1976).

Adjacent rock formations are frequently extremely important in determining how karstic phenomena will develop on any particular carbonate rock. Their effect ranges from protection of directly underlying limestone from downward percolation of rainwater (e.g. through the presence of shale bands) to increased solution if the overlying rock is porous and admits acid waters (e.g. through the presence of quartzose sands and gravels). In addition, overlying non-carbonate strata may allow run-off to reach a carbonate area in an aggressive state ready to penetrate and corrode the limestone. They also provide clastic materials such as sand particles and pebbles to abrade underground cave systems. Thus the Namurian

Plate 19. Some of the most classic karst scenery in England is found in the Malham area of north west Yorkshire where there are limestone pavements, dry valleys, gorges and dolines. (Michael J. Stead)

shale capping to carbonate rocks in the Ingleborough district contributes to the classic karst development in the area, while in the Mendips the Old Red Sandstone cores of the Mendip anticlines provide an important source of run-off for cave and depression development.

Although the Carboniferous Limestones are pre-eminent as hosts for karstic development in England and Wales, within them there are notable differences in lithology which serve to control the forms of

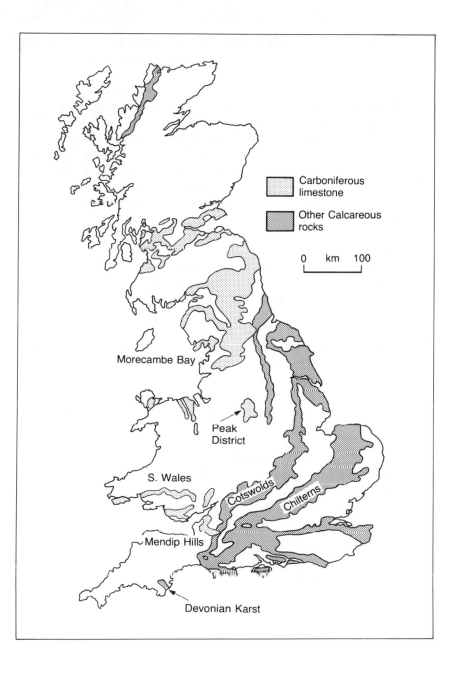

Figure 7.1 Areas of carbonate rocks in Britain.

such development. Sweeting (1970, p. 235), after thin-section study of limestones from northern England, made these generalizations:

> . . . the sparites tend to form the more massive cliffs. Streams, both surface and underground, are concentrated along the joint planes, and narrow canyon-like valleys and narrow cave-passages are formed. Large-scale collapse of the massive blocks is rare, and the sparites are less susceptible to frost than are the other types of limestone . . . Enclosed hollows formed in these massive beds tend to be deep and narrow and to weather back in steep walls by parallel retreat. The impressive steep walled chasms or potholes so characteristic of the upper beds of the Carboniferous Limestones in the Ingleborough district of Yorkshire tend to originate predominantly in these highly sparitic and massive beds. These sparry beds also give rise to the most important limestone pavements.

She contrasts the sparitic limestones with the biomicrites, which in northern England tend to give rise to lower cliffs or crags. She explains (p. 237): 'These rocks are more porous, and this together with the greater frequency of jointing means that they tend to be more soluble and easily weathered. Valleys are wider and shallower . . . Collapse is also important, and many of the larger collapsed caverns in NW Yorkshire are associated with the micrites and biomicrites.'
  In the classic Ingleborough district it is the Great Scar Limestone which tends to provide some of the finest karst landforms in Britain. Lying near the base of the Carboniferous this is a remarkably pure limestone c.150–180 m thick, which is well bedded and massive and gently dips northeast at 1–3°.

Current rates of limestone denudation

By analysing karstic drainage waters for chemistry and discharge it is possible to measure the rate at which denudation is currently occurring. This rate is conventionally expressed as $m^3 \ km^{-2} \ y^{-1}$ (table 7.1).
  The data are divided into three groups according to the age of limestone from which they are achieved: Carboniferous, Jurassic

Table 7.1 Rates of limestone solution in England and Wales

Location	Rock type	Rate $(m^3 \, km^{-2} \, y^{-1})$	Source
North Cotswolds	Jurassic	35	Goudie (1967)
Corallian Scarp, Oxfordshire	Jurassic	74	Paterson (1970)
North Oxford Heights	Jurassic	60–66	Paterson (1970)
Berkshire Downs	Chalk	67	Paterson (1970)
East Anglia	Chalk	25	Perrin (1955)
Dorset	Chalk	50	Sperling et al. (1977)
South Downs	Chalk	55–65	Williams and Robinson (1983)
E Yorkshire	Chalk	39	Pitman (1986)
South Wales	Carboniferous	16	Groom and Williams (1965)
Mendips (E)	Carboniferous	50–102	Drew (in Smith 1975b)
Mendips (S central)	Carboniferous	81	Atkinson (in Smith 1975b)
Mendips (Cheddar)	Carboniferous	38–45	Ford (in Smith 1975b)
Mendips (Cheddar)	Carboniferous	23–29	Newson (in Smith 1975b)
Gower	Carboniferous	30	Chambers (1983)
Derbyshire	Carboniferous	75–83	Pitty (1968)
NW Yorkshire	Carboniferous	83	Sweeting (1966)

and Cretaceous (Chalk). From this grouping it is apparent that while there is a considerable range in values for different locations (from 16 in South Wales to over 100 in the Mendips) there is no great difference between the three rock classes. The mean value for the Carboniferous areas is 55, for the Jurassic areas 59 and for the Chalk 50. This indicates that the very different forms of karstic solution between the three rock classes are not the result of very different rates of denudation.

The data also indicate that present rates of limestone denudation in Britain are not insignificant. Indeed, Smith and Atkinson (1976) have analysed comparable data from other parts of the world, and indicate mean rates of solutional denudation in the tropics of $45.5 \, m^3 \, km^{-2} \, y^{-1}$; in temperate regions of 56.9; and for arctic and alpine areas of 61.8. Clearly, the British values are remarkably similar to those in many other parts of the world.

While it is plainly dangerous to extrapolate these data because of the effects of such factors as climatic change, they do indicate that substantial landscape modification may have occurred even within the span of the Pleistocene through solutional attack alone. A rate of denudation of 50 m^3 km^{-2} y^{-1} is equivalent to a rate of landsurface lowering of around 80 m during the 1.6 million years of the Pleistocene.

Long-term rates of drainage incision

An alternative method of trying to assess the rate of landform evolution and the age of limestone features is to undertake speleothem dating, using isotopic means, of which the Uranium–Thorium method has been the most useful.

The principle behind this method is that many areas have high-level, abandoned cave passages in which former streams drained to valley floor levels now well above the present one. Thus, by dating speleothem deposits in the caves – deposits which could not have formed while the cave was still active – the date when the caves were abandoned by their streams can be assessed. This means that rates of stream downcutting can be obtained by comparing dates with levels, with the vertical distance between dated cave levels indicating the rate of downcutting (Trudgill, 1985).

In north-west Yorkshire (Gascoyne et al., 1983) large, high-level cave systems, such as Ease Gill Caverns, the West Kingsdale caves, and Gaping Gill–Ingleborough Cave have been found to have ages beyond the limit of the $^{230}Th/^{234}U$ dating method (i.e. greater than 350,000 years). Such caves are at up to 100 m above present levels, and there is an intermediate stage, some 50 m above present levels, which dates to about 250,000 years BP. It is calculated that mean maximum downcutting rates in limestone channel beds in the Craven Area are around 20–50 mm 1000 y^{-1}.

Comparable work has been done in Derbyshire, where there are also caves with dates greater than 350,000 years (Ford et al., (1983). Indeed, Rowe (1988) has proposed, on the basis of both uranium series and palaeomagnetic dating, that some speleothems in the Manifold Valley may be as much as 1.8–2.0 million years old. He calculates maximum incision rates of 55 mm 1000 y^{-1} for the Manifold, and of 63 mm 1000 y^{-1} for the Cresswell Crags Gorge.

Likewise, in the Mendips, karst scenery was well developed before 350,000 years BP (Atkinson et al., 1978). There has been a decline in local base level by drainage incision of some 70 m, a rate of approximately 20 mm 1000 y^{-1}. This is broadly the same as the rate for Craven.

Ancient karsts

Although glacial erosion, periglacial run-off and current solutional activity indicate that a substantial modification of karstic landscapes has taken place in the Quaternary, there is some evidence that elements of Karst landscapes may be extremely old (Bull, 1980; Douglas, 1987). Various phases of active karstification have been identified (Wright, 1986):

1 Some karstification took place during deposition of Lower Carboniferous beds (Intra-Carboniferous palaeokarst).
2 Considerable erosion may have occurred at the end of carbonate deposition, before re-submergence and deposition of Namurian sediments (e.g. Millstone Grit).
3 In the Trias and Jurassic, areas like the Mendips had their Carboniferous beds exposed and were subjected to erosion under arid conditions. Likewise, the sub-Triassic landscape of South Glamorgan is one of a Carboniferous Limestone island surrounded by lithified fossil scree slopes.
4 In the Tertiary, some solution depressions developed which became filled with sands, clays, lignite etc. (e.g. in North West Wales – Walsh and Brown, 1971). A notable example is the preservation of the Neogene Brassington Formation in isolated hollows in the Derbyshire Peak District (Ford, 1984). Corbel (1957) suggested that certain conical hills in the Morecambe Bay area (e.g. Arnside Knott) were relict Tertiary tropical Karst (though see Vincent, 1985, for an alternative explanation).
5 Pre-Devensian erosion and solution is made evident by the considerable age of some cave systems as indicated by isotopic dating (see p. 220). Furthermore, the size of certain doline and polje-like features precludes a modern formation, and they often contain Devensian or pre-Devensian sediments.

Closed depressions

Closed depressions are common in British limestone areas and occur on the Chalk and Jurassic outcrops (see pp. 232 and 235), as

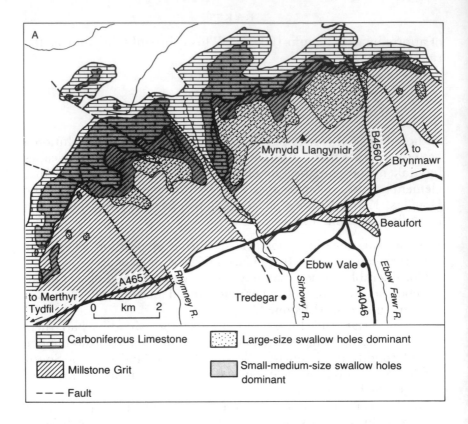

A

	Carboniferous Limestone		Large-size swallow holes dominant
	Millstone Grit		Small-medium-size swallow holes dominant
- - -	Fault		

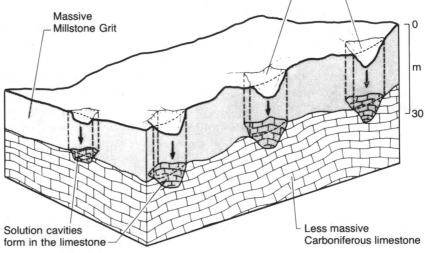

B

Depressions form as grit collapses into solution cavities

Massive Millstone Grit

Solution cavities form in the limestone

Less massive Carboniferous limestone

well as on the Carboniferous. They have a large variety of sizes and forms.

The largest class of depression, with flat floors and cliffed margins are *poljes*, although British examples are less well developed than the classic Yugoslavic forms. Vincent (1985) suggests that poljes exist in the Silverdale Peninsula area to the east of Morecambe Bay, good examples being New Barns Moss and Silverdale Moss.

Smaller, circular depressions, called dolines are more widespread, and have a variety of local names, including swallow-hole, swallet and sink-hole. They also have a variety of origins. *Solution dolines* develop at some favourable point, such as a joint intersection, where surface solution is concentrated. The solutes and some of the insoluble residues are evacuated down solution-widened joints and bedding-planes. Such a focus of solution and downward percolation gathers drainage to itself and an initial depression becomes enlarged. This is the favoured mechanism for the dolines of the Mendips (Smith, 1975a). It is also the mechanism favoured by Clayton (1981) for a series of about 20 closed hollows north-east of Malham. These are often one-third to one kilometre across and approach 100 m in depth. O'Connor et al (1974) suggest that as the Great Scar Limestone is so massive and strong it is unreasonable to postulate collapse as responsible for any major features in the area.

A second type of doline, the *collapse doline*, is formed by the collapse of the roof of a cave by underground solution. Several dolines which are clearly of collapse origin occur in north-west Yorkshire: Gavel Pot on Leck Fell and Hill Pot on Peny-y-Ghent (Gunn, 1985).

Another related type of doline is the *subjacent karst doline*. These are dolines found in other rock formations overlying karst rocks. A classic area for these is the South Wales Coalfield, (figure 7.2A), where collapse features attain their maximum development on the Millstone Grit (Thomas, 1974). The Grit has the mechanical strength to form a relatively stable roof to caverns formed in the underlying limestone; these caverns are therefore able to grow to a considerable size before collapse occurs. The suggested mechanism for these hollows is shown in figure 7.2B. Interstratal solution collapse on this scale, producing some thousands of dolines over

Figure 7.2 The distribution of dolines in South Wales (A) and the mechanism of formation by solutional collapse beneath a cover of Millstone Grit (B).

about 60 km length of outcrop, some with diameters in excess of 100 m, has not been found elsewhere in Britain and the reason (Ford, 1984, p. 260) 'is probably because it is only in South Wales that the basal Millstone Grit beds are coarse permeable sandstones which allow acid waters to percolate directly into the limestone. Elsewhere in Britain thick shales intervene.' The dolines attain a density of 50–75 features per hectare (Thomas, 1963) and may occur on Millstone Grit thicknesses of over 150 m.

Subsidence dolines occur where thick soils or superficial deposits (e.g. glacial drift or loess) cover karst rocks. Dolines can then develop through spasmodic subsidence and more continuous piping of these materials into solution pipes and widened joints (Jennings, 1985). The 'shakeholes' of north-west England are just such features, and are cone-shaped depressions often about 3 m deep and 8–11 m in diameter (Gunn, 1985).

Swallow holes are another type of limestone depression, and rivers disappear into them. In north-west Yorkshire they are called 'potholes', and this is the classic area in Britain for their development. It is here that one has the necessary conditions for their formation: a massive, sparry limestone with strong vertical jointing; considerable local relief with the karstic area perched high above the surrounding non-karstic rocks; a bounding fault system; a recent history of glaciation and meltwater run-off; a high rainfall; and inputs of water from non-karstic shale exposures (e.g. from Ingleborough). Features vary from deep narrow shafts (e.g. Long Kin West pothole on the southern slopes of Ingleborough) to large tunnels and associated caverns beneath (e.g. Gaping Gill).

Gorges

Although on a global basis gorges are found in practically all rocks, they are more frequent and bolder in karstic limestone areas than on any other rock type. Britain is no exception, possessing spectacular examples in areas like the Mendips and north-west Yorkshire. The essential mechanism in this prevalence of gorges in limestone areas is the failure of slope processes to flare back the valley sides. As Jennings (1985, p. 88) explained, 'Marked infiltration and reduced runoff minimize slope wash and many kinds of mass movement that tend to widen valleys and moderate their steepness.' Although this may be the common factor linking

Plate 20. Goredale Scar, in north-west Yorkshire, is developed in the more massive facies of the Carboniferous limestone. (Michael J. Stead)

the development of many different gorges, there is a wide range of explanations to explain the initial development of a steep-sided stream channel in limestone areas.

One explanation for limestone gorges, including Daniel Defoe's 'deep frightful chasm' of Mendip, the Cheddar Gorge, was that they originated through the collapse of an underground cavern or caverns. Smith (1975a) provides a historical analysis of the growth of this concept; Ford and Stanton (1968) demolish it. The concept is fundamentally unsound because of the disparity in size between the gorges and known caverns in Britain. In addition, there is

considerable evidence that features like Cheddar Gorge were cut by
Pleistocene periglacial run-off: the spurs that make up the side of
the gorge interlock, just as in a normal river valley; the gorge heads
into a well-developed drainage network; run-off can be effective in
big storms even today; and the gorges have cut down vertically
through pre-existing cave passages.

The Winnats Gorge in Derbyshire provides an illustration of the
complexity of gorge formation. It is a culmination of a sequence of
superimposed developmental stages (Ford, 1986). Initially it was a
lower Carboniferous (?) inter-reef channel dating to the late Asbian
and early Brigantian. It was then deepened during the pre-

Plate 21. Cheddar Gorge, on the flanks of the Mendip Hills, has
sometimes been thought of as a collapse feature, but in reality is a dry
valley that heads backward into a well marked valley system on the
Mendip plateau. (Cambridge University Collection: copyright reserved)

Plate 22. The Winnats Gorge, near Castleton in Derbyshire, has a complex history, but may in part be an exhumed inter-reef channel dating back to the Lower Carboniferous. (Author)

Namurian (Upper Carboniferous) palaeokarstic phase, before being filled in with Namurian shale. It probably remained filled until ice-margin run-off in late Wolstonian times scoured it out. It was further trimmed during Devensian periglacial phases.

Other gorges result from the superimposition of a drainage system on to the underlying limestone from a cover of rocks that has since been stripped off. Although the age of the cover rocks responsible is not agreed, this is the preferred explanation of the major limestone gorges of the Derbyshire Peak District, such as the Derwent's gorge in Matlock and Dovedale. However, once the drainage was superimposed, any prospect of uniclinal shifting away from the limestone was halted by the presence of resistant reef knolls within the Carboniferous Limestone (Ford and Burek, 1976).

Limestone pavements

Limestone pavements are exposed areas of bare limestone, both flat and sloping, which are often embroidered with an array of microforms produced dominantly by solutional activity. In the continental literature, the microforms are called *karren* or *lapiés*. In Yorkshire the terms clint and grike are employed, the clints being the blocks of limestone that constitute the paving and grikes the open crevices isolating the individual clints.

Plate 23. Most limestone pavements in Britain have been created by glacial stripping. At Norber, non-calcareous erratics rest on pedestals on the underlying Carboniferous Limestone. (Author)

A crucial issue is to explain what stripped a particular limestone area to give an exposed pavement upon which solutional activity could subsequently work. The most persuasive explanation is that the stripping was produced by glaciation and Pigott (1965) noted that bare limestone surfaces in Britain are characteristic of limestone districts within the limits of Newer Drift (Devensian). He explained the absence of exposed pavements on the Derbyshire limestone plateau as a consequence of a prolonged period of weathering and an absence of Devensian glacial stripping. This view was supported by Williams (1966, pp. 158–9):

> In the British Isles the largest exposures of limestone pavements are in the north Clare–south Galway district of Western Ireland, and in the Ingleborough locality of the Pennines. Patches of pavement also occur around the Lake District, in Sutherlandshire, and in South Wales. In

central England, the limestone uplands of Derbyshire do
not possess pavements, and father south, they are also
absent on the Mendip Hills . . . All areas possessing
extensive limestone pavements were glaciated in the last
glaciation . . . Glacial scouring provides the means of
wholesale removal of any loose or fractured material that
may formerly have lain on the surface . . . By contrast,
just south of the limit of the last glaciation in Britain, the
limestones are commonly buried by loamy soils to a depth
of several feet, and bedrock outcrops are uncommon,
except on steep valley sides.

The limited development of pavements on the Carboniferous
Limestone outcrop of South Wales (214 patches with a total area of
only 73 ha) has been attributed by Thomas (1970a) to the extensive mantling of the solid formations by boulder clay and
collapsed masses of Millstone Grit, and to the fact that in this area
the Carboniferous Limestone for the most part has few massive
beds (a contrast to the Great Scar Limestone of Craven).

Another important control of pavement distribution is limestone
lithology and structure. For good pavements to form the limestone
needs to be pure otherwise large amounts of insoluble residue will
mantle the bedrock; it needs to be resistant to frost attack; it needs
to be well bedded; and it must not be too soft or highly fractured.
For this reason limestone pavements are restricted to portions of
the Carboniferous Limestone – they are absent from Jurassic and
Cretaceous limestones.

In post-Devensian times the pavements have been subjected to
modification by solutional processes, probably suffering a mean
overall lowering of 0.4 m in the last 10,000 years (Thomas, 1970a).
A distinction can be made between forms with rounded divides,
which have formed under a peat or soil cover (the *rundkarren* of
German workers), and those angular forms produced directly by
the dissolving power of rain (the *rillenkarren* of German workers).
Examples of miscellaneous forms are illustrated in figure 7.3. The
nature of clints and grikes has been studied by Goldie (1973). She
provides data on the sizes of the features and demonstrates some
variability within the Craven area of north west Yorkshire
(table 7.2). The grikes average about 1 m deep and 0.2 m across,
whereas the clint blocks average just under 3 m long and just over
1 m in width.

Data on the nature and origin of small basins or pavements,

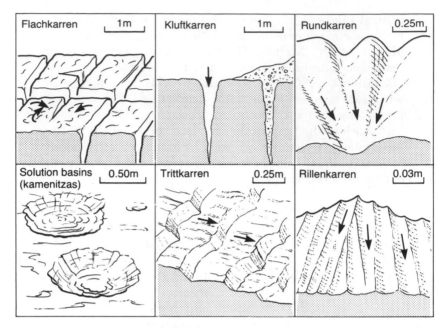

Figure 7.3 Selected karren types from the karst region of Morecambe Bay (after Vincent, 1985, figure 9.11).

called *Kamenitzas*, is given by Rose and Vincent (1986).

Lewis (1983) has undertaken morphometric work on the pavements of the South Wales coalfield and found that grikes averaged 14 cm wide and depths between 11 and 50 cm. The mean clint length was 88 cm and the mean width 49 cm. They are thus markedly smaller than the Craven examples. Another example of morphometric work is that undertaken on Cumbrian pavements in the vicinity of Morecambe Bay (Goldie, 1981). Some of these pavements are steeply dipping (e.g. Hutton Roof Crags) and have narrow clints and grikes, but for the most part their dimensions are more like those of Craven than of South Wales.

Because of the picturesque nature of these various surface embellishments, limestone pavements have been subject to widespread damage and wholesale removal by humans (Goldie, 1986).

Table 7.2 Mean dimensions of clints and grikes

Location	Grikes		Clints	
	Depth (m)	Width (m)	Width (m)	Length (m)
Craven				
Ingleborough				
Twistleton Scar	1.13	0.18	1.19	2.19
Scales Moor	1.06	0.22	1.49	3.17
Scar Close	1.18	0.17	1.95	4.04
Southern Scales	1.17	0.20	1.69	5.67
Raven Scar	0.49	0.14	0.53	1.68
White Scars	0.67	0.26	1.19	2.13
Clapdale Scars	0.69	0.08	0.36	0.86
Borrins Moor	1.32	0.20	1.75	3.97
Malham				
Malham Cove	1.00	0.15	1.06	2.94
Back Pasture	0.81	0.21	0.55	2.64
Pen-y-Ghent	0.85	0.22	0.82	1.92
Wharfedale				
Oughtershaw	0.79	0.21	0.99	2.77
Halton	0.86	0.14	0.51	1.33
Blue Scar	1.05	0.13	0.48	1.89
Low Far Moor	0.66	0.09	0.93	1.51
Threshfield	0.59	0.26	0.84	2.09
Conistone	0.70	0.12	0.86	1.84
Grass Wood	0.81	0.19	0.94	1.60
Mean	0.98	0.20	1.23	2.83
Cumbria				
Holme Park Fell	1.05	0.13	2.32	3.15
Hutton Roof Crags	0.17	0.15	0.90	2.30
Farleton Fell	1.21	0.17	1.05	2.75
Newbiggin Crags	1.00	0.20	1.59	2.20
Mean	0.60	0.16	1.47	2.60
South Wales				
Mean	0.11–0.50	0.14	0.49	0.88

Source: data in Goldie (1973, 1981) and Lewis (1983)

Plate 24. Bedding inclination and bedding thickness are important controls of limestone pavement lithology. This pavement at Hutton Roof Crags, in Lancashire, has developed on steeply dipping beds, and displays well developed clints, grikes and rills. (Author)

Karstic phenomena on the Chalk

The Chalk outcrops of southern and eastern England are often considered to lack many characteristic features of karst landscapes, but this is a serious misconception, for dolines and swallow-holes are widespread.

The most impressive Chalk dolines are those of Dorset (Sperling et al., 1977) concentrated in heathland areas between Dorchester and Bere Regis, notably on Puddletown Heath, Southover Heath and Affpuddle Heath. The hollows attain extremely high densities: 99 km^{-2} on Southover Heath and 157 km^{-2} on Puddletown Heath. The largest hollow, Culpepper's Dish, has a diameter of 86 m and a depth of over 21 m. Collapses of the ground into hollows are still taking place today.

Dolines occur widely on other parts of the English chalklands, including the Yorkshire Wolds near Raisthorpe (Lewin 1969); some quantitative data are provided by Edmonds (1983; figure 7.4). The

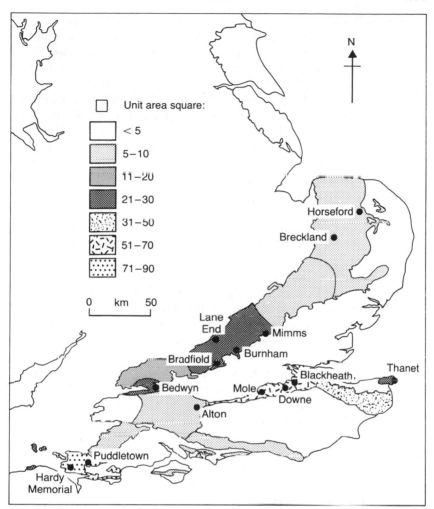

Figure 7.4 The density of solution features (dolines) per 100 km²
from the Chalk of England (modified from Edmonds, 1983, figure 1
and Sperling et al., 1977, figure 13).

lowest-density regions are Salisbury Plain, Yorkshire and Lincoln-
shire. The higher-density regions, besides Dorset, are the Chiltern
Hills, Pewsey area, Kent Downs, and Surrey and West Kent
Downs.

Some East Anglian hollows, a number of which are not
solutional (Prince, 1962), closely resemble the Dorset hollows and

modern collapses have been recorded (Whitaker, 1921, pp. 41–3). They occur mainly in the Breckland, and are particularly numerous north and east of Thetford, especially near Croxton, East Wretham and Roundham Junction. The superficial materials over the Chalk are miscellaneous types of Quaternary drift.

There are some dolines on river terraces and valley sides in the Mimms Valley area of Hertfordshire (Kirkaldy, 1950, p. 220) where London Clay, Reading Beds and some glacial deposits overlie Chalk. Hare (1947, p. 327) has mapped the distribution of 'swallets' on the Chalk and Reading Beds near Burnham Beeches, Buckinghamshire, and has shown that while most occur near the Reading Beds–Chalk margin, or where the Reading Beds are thin, some appear to have developed on up to 12 m of Reading Beds. Another famous example of chalk swallow-holes are those that conduct the flow of the River Mole in Surrey (Edmunds, 1944).

There appear to be two major controls on the distribution of solutional features on the Chalk. One of these is the lithology of the Chalk itself, and Edmonds (1983, pp. 264–5) has shown that there is a preferential development on the Upper Chalk, the purest, softest and most porous chalk, with the fewest restrictions to groundwater flow. However, the most important distributional control is the presence of Tertiary Beds or Quaternary superficial deposits on top of the Chalk (figure 7.5). Such deposits may concentrate water into particular locations on the Chalk surface, but their most important role is to provide a source of acidic water.

Solution of the Chalk appears to have been a long-continued process for while examples of modern collapses are legion, there are also examples of ancient dolines in the Chalk. For example, Gibbard et al. (1986) demonstrate, by an analysis of sediments within it, that a large doline near Denham in Buckinghamshire pre-dates the Hoxnian interglacial.

The local presence of fully developed solution fissures in parts of the Chalk outcrops has certain important hydrological implications, promoting rapid, turbulent subterranean flow over long distances. Atkinson and Smith (1974, p. 204) have drawn attention to such a situation in South Hampshire:

> . . . swallow holes within the Havant–Bedhampton catchment area occur only around the northern margins of the Tertiary outcrop, at sites where seepage and runoff has enhanced erosion by solution of the Chalk. It may be tentatively suggested that wherever the Chalk is capped

by relatively impermeable strata, perhaps including the clay-with-flints, [see below] zones of higher permeability than usual or fully developed solution-fissures are likely to occur. This would indicate a similar overall groundwater situation to that more commonly associated with massive limestones.

To the south of the zone stripped of superficial deposits by the various glacial advances, the Chalklands are mantled over extensive areas by a group of deposits called clay-with-flints (figure 7.6). They are highly variable in composition ranging 'from heavy reddish brown clays with large unworn flint nodules to almost stoneless yellow or white sands, yellowish to reddish brown silt loams, brightly mottled (red, lilac, green and white) stoneless clays, and beds of rounded flint pebbles' (Catt, 1986b, p. 151). Early English geologists tended to regard them as the insoluble residue, left after a long period of dissolution and weathering of the Chalk. However, although some of the constituent material of clay-with-flints may have been derived from this source, it is not an adequate explanation of the variability of the material nor of the presence of miscellaneous types of clay, sand and flint shape. Much of it is probably derived and reworked from Palaeogene beds, as Jukes-Brown (1906) so astutely recognized.

Karstic phenomena in Jurassic limestones

The Jurassic limestone backbone of England displays a limited amount of karstic development, but in general it fails to reach the importance of that on the Carboniferous limestone or that on the Cretaceous Chalk.

Where karstic phenomena do occur it is either because of a cover of superficial material (as in Lincolnshire) or because of the presence of non-carbonate impermeable marls and clays within a limestone sequence (as in parts of the Cotswolds).

The Jurassic Lincolnshire Limestone between Grantham and Stamford (figure 7.7) illustrates the first of these situations, possessing many sink-holes. Hindley (1965, p. 458) explains their existence and distribution as follows:

> The most striking conclusion is that sinks do not occur on
> the drift-free Lincolnshire Limestone, but rather towards

the thinning edges of the boulder clay or, especially south of Stretton, of the Upper Estuarine Clays where they rest on top of the Lincolnshire Limestone. The reason is simply that precipitation which falls upon the exposed limestone is too readily absorbed by normal percolation for streams to form; and without some such concentration of flow no concentrated solution of the limestone can take place, and hence no sinks form. The boulder clay cover provides the necessary catchment and concentration of surface flow, and as the clay thins the water finds its way through to the underlying limestone and opens up swallets.

The Oolitic limestones of the Cotswolds form an impressive cuesta across which run the courses of the tributaries of the Thames, such

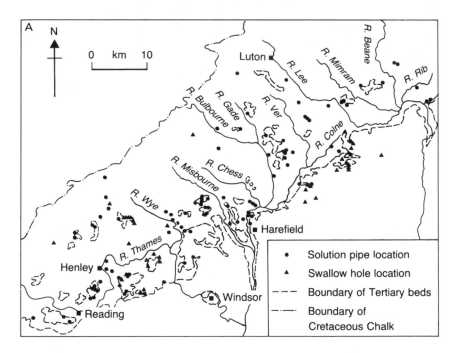

Figure 7.5 The distribution of solution features on the Chalk: (A) swallow-holes in the Chilterns (from Edmonds, 1983, figure 2); (B) Dolines in the Dorset heathlands (from Sperling *et al.* 1977, figure 11). Note in both cases, the association with the presence of Tertiary beds over the Chalk.

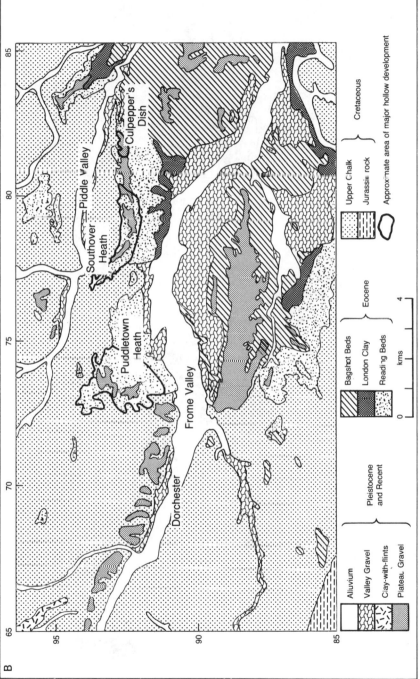

B

Pleistocene
and Recent

Alluvium

Valley Gravel

Clay-with-flints

Plateau Gravel

Eocene

Bagshot Beds

London Clay

Reading Beds

Cretaceous

Upper Chalk

Jurassic rock

Approximate area of major hollow development

kms

0   4

Dorchester

Frome Valley

Puddletown
Heath

Southover
Heath

Piddle Valley

Culpepper's
Dish

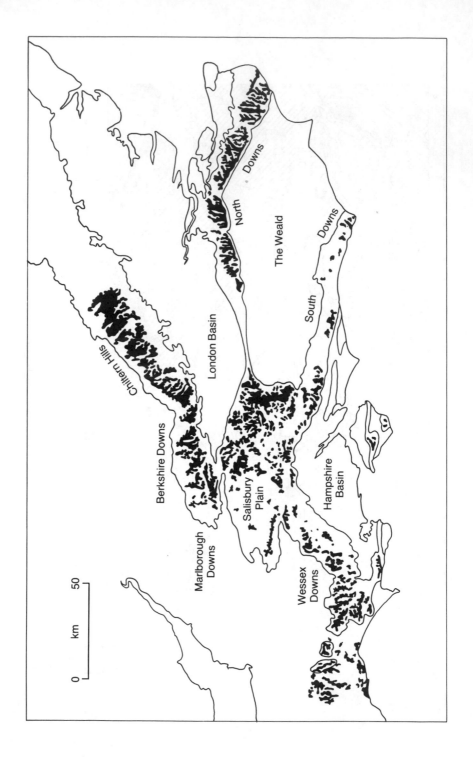

as the Coln, Churn, Leach, Windrush etc. Some of these streams suffer from discharge diminution as they flow from non-carbonate beds on to limestones; some go underground, at least seasonally, for part of their length. The most notable example is the River Leach (Goudie et al., 1980), whose long underground stretch coincides approximately with the surface outcrop of the Great Oolite.

Karstic phenomena on non-carbonate rocks

Karstic phenomena in Britain are not restricted to carbonate rocks. There are other soluble rock types in which miscellaneous natural depressions occur. Two main groups can be recognized: those developed in Triassic rock salt beds and those in Permian gypsum.

The Triassic rock salt beds of Cheshire have been dissolved to give both crater-like depressions (10–200 m in diameter) and linear hollows (up to 240 m in width and 8 km in length). Although the rate of solution and collapse has been increased by human activities connected with brine pumping and salt mining (see p. 308) there is no question that some of the features are both old and natural. As Howell and Jenkins (1976) point out, post-glacial peats and sediments are present in many hollows, and medieval land boundaries reflect the presence of ancient subsidence lines.

'Singular pits and depressions' (Tute, 1870, p. 3) occur in Permian gypsum country north east of Ripon, Yorkshire, and 'In form, they are generally crater-like hollows, from fifty to a hundred feet across, and sometimes, though very rarely, large and deep perpendicular shafts. There are about thirty or forty of these regularly-formed depressions. But a large portion of the district has been thrown into a succession of little hills and valleys in consequence of other irregular subsidences.' Tute gives details of various collapses, notably in 1836 and 1860, while James et al. (1981) demonstrate how gypsum cliffs on the River Ure are being sapped by current solutional activity.

The solutional origin of many holes in non-carbonate terrains is, however, far from clear, for depressions can result from a whole

Figure 7.6 The distribution of plateau drift and clay-with-flints in southern England (based on 1:250,000 maps of the Soil Survey of England and Wales). The Chalk outcrop is outlined by a thick solid line.

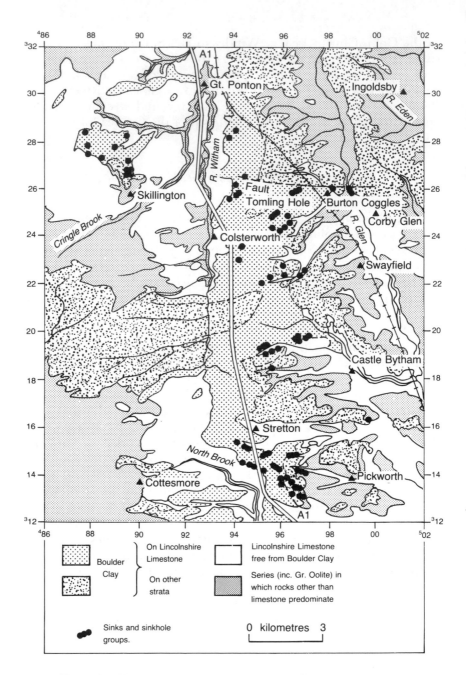

Figure 7.7 The distribution of sinkholes on the Lincolnshire limestone between Grantham and Stamford (after Hindley, 1965, figure 1).

array of non-karstic processes. So, for example, although Bennett
(1908) identified 'solution-subsidence valleys and swallow-holes'
within the Hythe Beds of Kent, which consist of sandstones,
limestones, sands and chert, it is likely that some of the Hythe Bed
depressions result from cambering (see pp. 96–101). Others appear
to be true swallow-holes (e.g. Langley Hole), and the Loose Valley
has a dry gorge section known as Boughton Quarries (Robinson
and Williams, 1984).

Limestone tufas

Although limestone solution is the dominant karstic process in
Britain, there are examples of where landforms have been created
by the deposition of calcium carbonate in the form of tufa.
Pentecost (1978) provides a map of the distribution of active and
fossil tufa deposits in Britain; this forms the basis of figure 7.8. Tufa
deposits occur more widely than this, however, and many *Memoirs
of the British Geological Survey* give locational details.

Tufa is a soft, porous calcareous rock that forms in springs,
waterfalls and lakes. Deposition is a chemical process, and typically
occurs where subterranean waters, supersaturated with calcite and
rich in carbon dioxide, issue onto the surface (Pentecost, 1981).
However, certain mosses and algae contribute to the depositional
process and mosses, liverworts, reeds, etc. help to provide an
open framework for the material.

There are some major tufa-deposition streams in the Malham
area, and Gordale beck is 'probably the best example of a tufa-
depositing stream in the British Isles' (Pentecost, 1981, p. 367).
Landforms include tufa dams, mounds and waterfall curtains.
Under favourable conditions growth rates may be between 0.2 and
1.6 mm per year. Other significant tufa landforms are the ancient
barrage tufa deposits at Caerwys in the Wheeler Valley of North
Wales (Pedley, 1987).

There is some evidence that tufa deposition may have been more
prevalent and widespread at certain times in the past, with rapid or
more extensive deposition prior to or during the Atlantic Period of
the Holocene (*c.*6,000–8,000 years BP); (Pentecost and Lord, 1988).

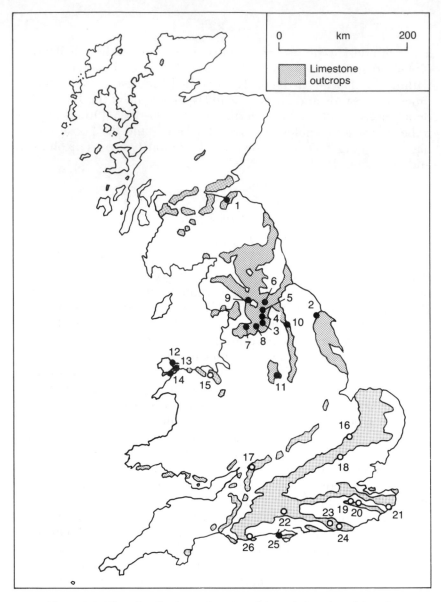

Figure 7.8 The distribution of active (●) and inactive (○) tufa-depositing sites from various surces in Pentecost (1978, figure 1):

1 = Pentland Hills	2 = Helmesley
3 = Gordale Beck	4 = Malham Tarn
5 = Waterfall Beck	6 = Kettlewell
7 = Clapham Beck	8 = Howgill
9 = Aysgarth	10 = Knaresborough
11 = Matlock Bath	12 = Penrhyn Glas
13 = Penmon	14 = Menai Straits
15 = Caerwys	16 = Bourne Brook
17 = Durley	18 = Harpenden
19 = Plaxtol	20 = Wateringbury
21 = Dover	22 = Helen River
23 = Clayton	24 = Lewes
25 = Totland Bay	26 = Blastenwell

8

THE COASTLINE

Introduction

The coastline of England and Wales is highly diverse. As the greatest exponent of its virtues, character and origin, J. A. Steers, put it (1960, p. 1):

> In continuous stretches like America, Russia, or Australia and, of course, in many smaller regions it is possible to travel far without noting any great change in scenery. But the British Isles possess both a very wide variety of scenery within a small area and also examples of rocks formed in nearly all the geological ages through which the earth has passed. The variety of rocks in England and Wales is largely responsible for the many types of coastline in the country.
>
> A recent estimate of the length of the coastline of England and Wales is 2,750 miles (c.4,400 km), and it is very rare to find the same kind of coastal scenery for more than ten or fifteen miles together.

The best regional guide to this diversity is provided in Steers's magisterial *The Coastline of England and Wales* (1948a). This chapter treats some of the major coastal phenomena in a systematic way.

Cliffs

Cliffs of different heights (figure 8.1) form the backdrop of long stretches of coastline. As Steers (1948a, pp. 65–6) has remarked:

> The cliff coasts of Great Britain should be reckoned one of the country's most treasured possessions. They are of every variety and form, including the great chalk promontories of Flamborough Head and Beachy Head, the granite masses of Land's End, the sandstones of St. Bees Head, the fine series of limestones of Tenby and Gower, the Old Red Sandstone and Carboniferous cliffs of western Pembrokeshire, the newer limestones, sandstones, and clays of Dorset, and the glacial cliffs of Holderness and Norfolk.

It is clear from this statement that lithology and structure are major controls of cliff form. Vertical cliffs, for example, often form in almost horizontal beds, especially where the bedding is massive and the composition relatively homogeneous, as in the chalk cliffs of Sussex and north-west Norfolk. If the bedding is thinner and there are frequent alternations of different lithologies (e.g. the Lias Cliffs near Lyme Regis) the form may be slightly more complex, although vertical sections may still occur as retreat may be rapid as a result of the undercutting of softer beds. Cliffs with beds dipping seaward tend to have a profile which is broadly coincident with the plane of the dip; cliffs with beds dipping inland are seldom precipitous because undercutting cannot readily produce cliff falls. Verticality is not invariably due to structure, however. Where wave action is dominant and the products of cliff collapse are easily removed, vertical cliffs will occur. Cliffs formed in unconsolidated glacial drifts (e.g. Holderness and parts of East Anglia) are examples of this type.

   Composite cliff forms may occur as a result of differential erosion brought about by the juxtaposition of markedly different lithological types (Trenhaile, 1987, p. 184). In Purbeck, Dorset, for example, Purbeck Limestones weather relatively rapidly to give a low-angle concavity above vertical cliffs of Portland Stone, while at Bull Nose (west of Barry in South Wales) shale gives a concave, gulleyed slope above a vertical cliff cut in resistant limestone. The juxtaposition of different lithologies, particularly where the dip is

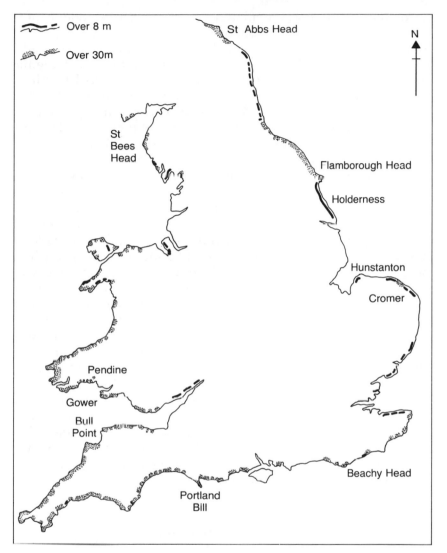

Figure 8.1 The distribution of cliffs on the coastline of England and Wales.

seaward, will lead to cliffs where landslipping creates complex forms with slipped masses forming undercliffs (see p. 253). Indeed, in locations which are relatively sheltered from wave attack subaerial processes such as mass movements may be dominant in shaping cliff profiles.

Another situation where subaerial processes become dominant is where marine action is reduced or even halted by such processes as accumulation of protective beaches or barriers. The sea cliffs gradually become degraded through time until they resemble comparable inland escarpment and valley-side slopes, as determined by the geotechnical properties of the rock outcrop (Bird, 1984, p. 75). A classic study of such slope decline through time is provided by Savigear's (1952) investigation of the cliffs at Pendine in South Wales, which were progressively abandoned as a spit grew eastwards (see p. 197).

In areas where cliff evolution is relatively slow, especially in areas of resistant lithology, cliff profiles may show the impress of a long history. It has to be remembered that in most areas the sea has only been shaping cliffs at the present level for about 6,000 years (i.e. at the end of the Flandrian transgression) and that sea-level has oscillated repeatedly during the Pleistocene. In general it is probable that cliff cutting occurred during interglacial high sea-levels, whereas degradation took place as a result of frost-weathering and solifluction during cold glacial periods with low sea-levels. Many cliffs are probably very old features which have only been modified to a modest degree by contemporary marine attack at their bases. That this is so is made evident by the presence of such phenomena as raised beaches, aprons of head and banks of cemented sandrock (dune sand), against cliffs (Orme, 1962). An idealized representation of a polycyclic and polygenetic coast, such as might be found in Devon and Cornwall, Gower and south Pembrokeshire is shown in figure 8.2. In areas that were glaciated, the modern cliffs may be seen to be being exhumed from beneath the till and other forms of drift as, for example, along parts of the Dyfed coast in mid-Wales (Wood, 1959).

Rates of cliff erosion

The intermittent nature of cliff retreat and possible inaccuracies and inconsistencies in cartographic portrayal make reliable estimation of long-term rates of cliff retreat somewhat problematic. Bearing this in mind it is possible to obtain a reasonably clear picture of those parts of the coastline of England and Wales where rapid retreat is occurring. Data on measured rates of retreat over approximately the last 100 years are presented in table 8.1.

Plate 25. The cliffs of the North Devon coast have a polygenetic origin, and current wave attack at their bases may be relatively limited in its effect. (Author)

As one might anticipate, the highest rates occur on certain rather specific stretches of coastline: the Yorkshire coast between Bridlington and Spurn Head; the Norfolk coast between Sheringham and Happisburgh; the Suffolk coast between Lowestoft and Southwold; the Essex coast at the Naze and Foulness; North Kent between Sheerness and Herne Bay; parts of the chalk coasts of Kent and Sussex; along the Sussex coastal plain to Selsey Bill; the south-west coast of the Isle of Wight; the Hampshire coast between Hurst Castle and Bournemouth; parts of the Dorset coast; Dawlish Warren; the Dee Estuary; and the vicinity of St. Bees Head on the Lancashire coast (Bird and May, 1976). It is therefore, with only a few exceptions, the lowland coasts of the east and south where rates of retreat are greatest.

The prime control is lithological, cliff retreat occurring rapidly either where there are unconsolidated glacial drifts or where there are relatively unresistant clays, shales, sands, or chalks. In such materials coast regression may occur at rates in excess of 100 m per century in exposed locations.

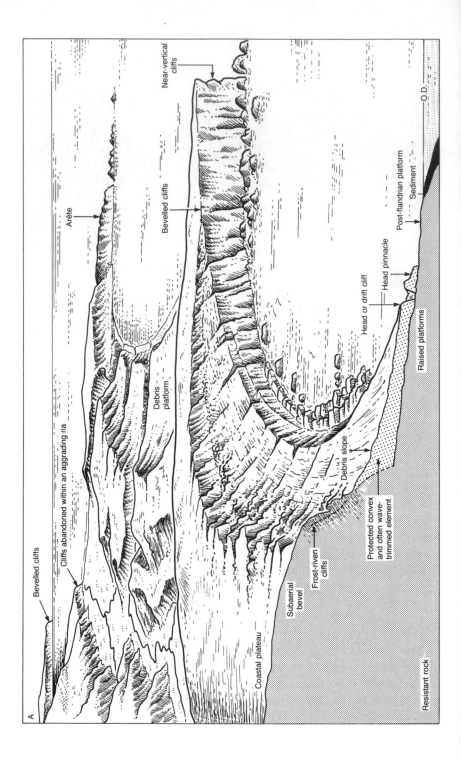

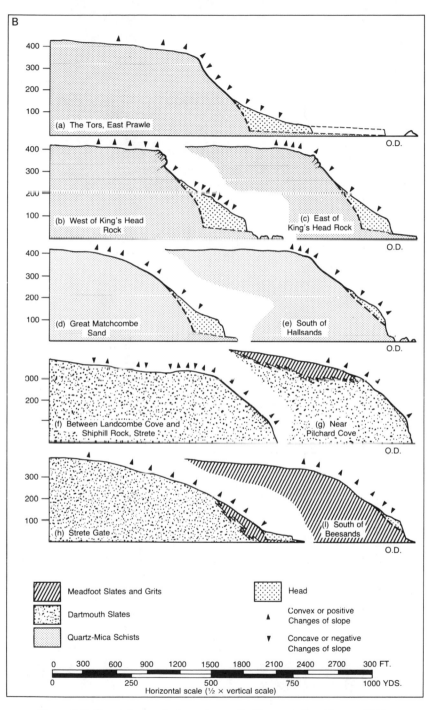

Figure 8.2 Coastal morphology in south Devon (after Orme, 1962, figures 1 and 2); (A) diagram of the modern and relict forms; (B) cliff profiles and head (solifluction) deposits.

Table 8.1 Examples of rapid coast retreat

Area	Geology of Cliff	Average rate retreat (m 100 y^{-1})	Source
North Yorkshire	Shale	9	Bird and May (1976)
North Yorkshire	Glacial drift	28	Bird and May (1976)
Holderness	Glacial drift	120	Bird and May (1976)
Norfolk			
Weybourne–Cromer	Glacial drift	42	Bird and May (1976)
Cromer–Mundesley	Glacial drift	96	Bird and May (1976)
Mundesley–Happisburgh	Glacial drift	88	Bird and May (1976)
Gratby–Caister	Glacial drift	83	Bird and May (1976)
Gorleston–Corton	Glacial drift	57	Bird and May (1976)
Pakefield–Kessingland	Glacial drift	105	Bird and May (1976)
The Naze (Essex)	Glacial drift, London Clay and Crag	11–88	Gray (1988b)
Kent			
Reculver	London Clay	68	May (1966)
N Isle of Sheppey	London Clay	96	
Isle of Thanet	Chalk	7–22	
St Margaret's Bay–Folkestone	Chalk	7–19	
Folkestone	Gault Clay	28	

East Sussex			Robinson and Williams (1983)
Peachhaven	Chalk	46	
Seaford Head	Chalk	126	
Birling Gap	Chalk	122	
Beachy Head	Chalk	106	
Ecclesbourne Glen	Hastings Beds (sandstone)	119	
Fairlight Glen	Hastings Beds (Clays)	143	
Cliff End	Hastings Beds (sandstone)	108	
Hampshire			May (1966)
Christchurch Bay (Highcliffe Castle)	Bracklesham Beds	3	
Christchurch Bay (Barton)	Barton Beds	58	
Christchurch Bay (Hordle)	Headon Beds	18	
Dorset			May (1966)
Ballard Down	Chalk	23	
Kimmeridge Bay	Kimmeridge Clay	39	
White Nothe–Hambury Tout	Chalk	21	
Ringstead	Kimmeridge Clay	41	
Furzy Cliff–Short Lake		37	
Isle of Wight			May (1966)
Cranmore	Hamstead Beds	61	
Newton River–Gurnard	Bembridge Beds	38	
Brighstone Bay	Wealden Beds	52	

Cliff mass movements

As we have already seen, on many cliffs subaerial mass movements contribute a great deal to morphology. Some major examples are listed in table 8.2. As can be seen, many of the major slips occur in relatively recent rocks, and in many examples clays and shales are involved. They are, therefore, relatively less common on the western and north-eastern coasts, though there are exceptions (e.g. those of St. Bees Head and Runswick Bay).

A useful distinction can be made between those mass movements which involve translational mudsliding and those which involve deep-seated rotational landsliding (Bromhead, 1979).

Table 8.2 Major examples of coastal landslips

Location	Lithologies	Reference
Folkestone Warren	Chalk/Gault Clay	Hutchinson et al. (1980)
Cromer, Norfolk	Pleistocene drift	
Essex coast (e.g. Hadleigh)	London Clay	Hutchinson and Gostelow (1976)
Isle of Sheppey, Kent	London Clay	Hutchinson (1968)
Stonebarrow, Dorset	Upper Greensand/Gault	Brunsden and Jones (1976)
Lizard, Cornwall	Devonian shales and sandstones	De Freitas (1972)
Brixham, Devon	Middle Devonian slates and shales	Derbyshire (1979)
Portland, Dorset	Portland limestone/ Kimmeridge Clay	Brunsden and Goudie (1981)
Axmouth–Lyme Regis, Devon	Upper Greensand/Gault and Liassic/Rhaetic	Pitts and Brunsden (1987), Pitts (1983)
Shanklin–Niton, Isle of Wight	Upper Greensand/ Gault Clay	Wright (1969)
Isle of Wight, North	Oligocene strata	Wright (1969)
Sandgate, Kent	Lower Greensand	Wright (1969)
Christchurch Bay, Hants	Barton Sand and Barton Clay	Barton et al. (1983)
Boscastle, Cornwall	Upper Carboniferous sandstones and shales	Arber (1911)
Runswick Bay, N Yorks	Deltaic Series Sandstones and Upper Lias shales	Rozier and Reeves (1979)
Fairlight Glen, E Sussex	Wealden (Hastings Beds) Farlight Clay and Ashdown Sand	Moore (1986)

The cliff on which translational mudsliding is dominant has a steep head slope and a much flatter front slope (figure 8.3). Debris from the head-slope, which is composed of inherently unstable material, falls or slides to feed the mudslide which occupies the flatter front-slope. If the lithology of parent rock is sufficiently variable there may be a more complex form, with spillage from upper mudslides supplementing that falling from lower head-slopes. The mudslides form a series of cirque-like semi-circular embayments and create tongues of debris that can extend a considerable distance on to the fronting beach. Bromhead (1979) gives as examples of translational mudslides those in the Thames Estuary which have developed on London Clay (e.g. on the Isle of Sheppey and between Herne Bay and Beltinge).

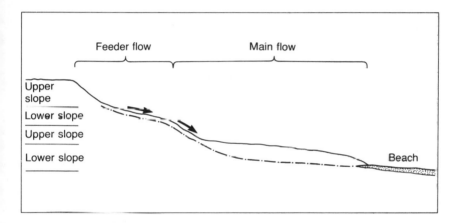

Figure 8.3 Section through a typical translational mudslide slope (from Bromhead, 1979, figure 1).

The second main type of mass movement on cliffs is the deep seated rotational landslide (figure 8.4). These arise when wave erosion oversteepens a cliff, so that failure occurs along a deep slip surface. This creates a landward scarp which is initially steep, but which degrades through time. Classic examples include Folkestone Warren (Kent), Stonebarrow (Dorset) and Blackgang (Isle of Wight). Where the cliff crest is stronger than the body of the slope because of the presence of a cap-rock the undercliff area may consist of multiple deep-seated rotational failures.

The nature of landsliding on cliffs is controlled in part by the rate

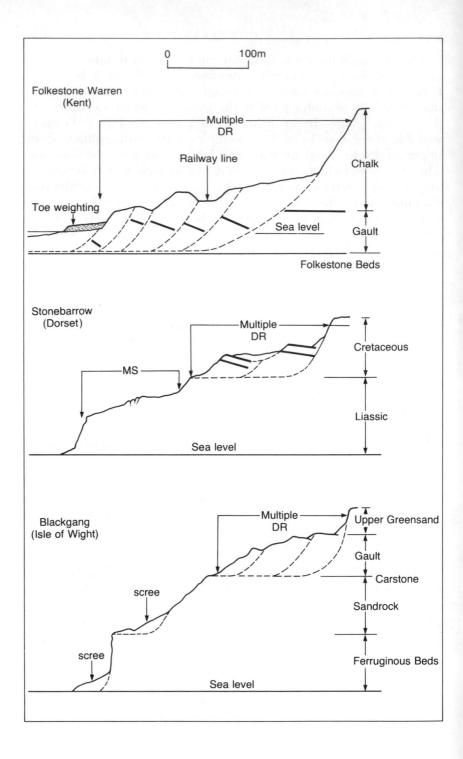

Plate 26. Part of the great landslip that afflicted the coast near Lyme
Regis in 1839. (Author)

of toe erosion, but also by the structural and stratigraphic
properties of the cliffs. Where the geological strike is perpendicular
to the coastline and the strata dip along the coast, the spatial
pattern of mass movement reflects the nature of the outcrop at sea-
level. Hutchinson (1983, p. 393) describes three different situations:

1 If competent strata occur at sea-level and the base of the
   overlying clay is above wave influence, cliff retreat is slow, the
   clay is drained from beneath, and only limited mudsliding
   occurs.
2 Where the dip of the beds brings the base of the clay within the
   upper quarter of the range of spring tides, moderate cliff-foot
   erosion is associated with active mudsliding.
3 Where the base of the clay layers occurs below sea level, because
   the recession of the cliff-foot is rapid, deep rotational slides take
   place.

Figure 8.4 Sketch sections through typical slopes with multiple
rotational slides (modified after Bromhead, 1979, figure 8).

Salt marshes

Salt marshes, the distribution of which is shown in figure 8.5, are vegetated tidal mudflats which form as fresh accumulations of fine

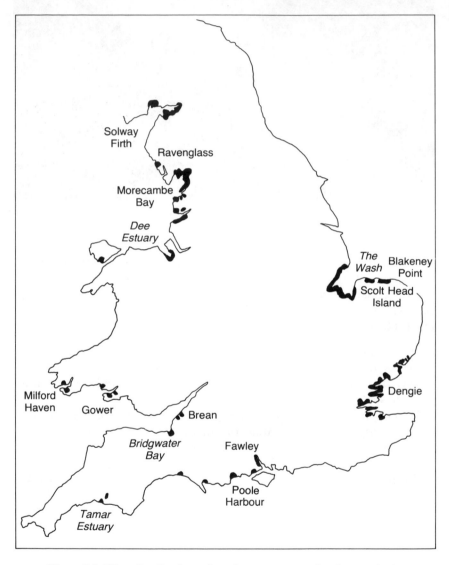

Figure 8.5 The distribution of major expanses of salt marsh in England and Wales.

sediment raise the level of the mudflat to a critical point above which it becomes exposed for long enough each day to allow vegetation colonization to occur. The vegetation then serves to trap more sediment, thus raising the marsh so that tidal inundation becomes less frequent and extensive. This in turn means that the environment for the marsh species has now altered so that different species, able to live on the higher marsh surface, can begin to colonize.

The normal British marshland vegetational succession is from *Salicornia* (marsh samphire), through *Halimione portulacoides* (sea purslane) – which often colonizes creek banks – and *Aster marlilmu* (sea aster) to *Puccenelia maritima* (salt marsh grass), *Limonium vulgare* (sea lavender) and *Armeria maritima* (sea pink). However, the sequence shows variability in different areas (see references in table 8.3).

Table 8.3 Select references to English and Welsh Salt Marshes

Location	Reference
Milford Haven	Dalby (1970)
Poole Harbour	Hubbard and Stebbings (1968)
Cefni, Anglesey	Packham and Liddle (1970)
Bridgwater Bay	Ranwell (1964)
Morecambe Bay	Gray (1972)
North Norfolk	Pethick (1980)
Solway Firth	Marshall (1962)
Brean (near Weston-super-Mare)	Page (1982)
Gibraltar Point, Lincs	Hartnall (1984)
Ravenglass, Cumbria	Carr and Blackley (1986)
Dengie, Essex	Harmsworth and Long (1986)
Dee Estuary	Marker (1967)
Severn Estuary	Allen and Rae (1988)

A comprehensive bibliography of British salt marshes is provided by Charman et al. (1986)

Rates of accretion on marsh surfaces can be quite rapid. In general rates of accretion are greatest on young marshes and gradually decline as marshes get higher, older and less subject to tidal inundation and sediment deposition. Pethick (1981) gives North Norfolk accretion rates on ten-year-old marshes of 1.7 cm y^{-1}, compared with less than 0.002 cm y^{-1} on marshes more than 500 years old. Other data for Norfolk are provided by Steers (1948b).

In the Severn estuary, Allen and Rae (1988) have established by means of archaeological, historical and sedimentological evidence that there has been around 1.22 m of vertical salt marsh accretion since the Roman period (of which 0.21 m has taken place since AD 1945).

In recent decades, however, the nature of some salt marshes, and the rate at which they accrete, has been transformed by a major vegetational change, namely the introduction of a salt marsh plant, *Spartina alterniflora*. This cord grass appears to have been introduced to Southampton Water by accident from the east coast of North America, possibly in shipping ballast. The crossing of this species with the native *Spartina maritima*, produced an invasive cord-grass of which there were two forms: *Spartina townsendii* and *Spartina anglica*. The latter is now the main species. It appeared first at Hythe on Southampton Water in 1870 and then spread rapidly to other salt marshes in Britain: partly because of natural spread and partly because of deliberate planting. For example, it reached Poole Harbour by *c.*1890 and the South Devon Estuaries in the 1940s and early 1950s. By 1971 it had reached Walney on the Cumbria Coast. Indeed, *Spartina* now stretches as far north as the Island of Harris in the west of Scotland, and to the Cromarty Firth in the east (Doody, 1984).

This plant has often been effective at excluding other species and also at trapping sediment. Rates of accretion can therefore become very high. Ranwell (1964) gives rates of 8–10 cm y^{-1} for Bridgwater Bay (Bristol Channel) and 2 cm y^{-1} for Keysworth Marsh (Poole Harbour). There is evidence that this has caused progressive silting of estuaries such as the Dee (Marker, 1967) and the Severn (Page, 1982).

However, for reasons that are not fully understood, *Spartina* marshes have frequently suffered from die-back, which has sometimes led to marsh recession. In the case of Poole Harbour and the Beaulieu Estuary, this may date back to the 1920s, but elsewhere on the south coast it has generally been rapid and extensive since about 1960 (Tubbs, 1984). The decline at Holes Bay in Poole Harbour is shown in figure 8.6. Equally dramatic has been the decline in the Wash, from 2,000 ha in 1958 to only 200 ha in 1973 (Doody, 1984). Among the hypotheses that have been put forward to explain die-back are the role of rising sea-level, pathogenic fungi, increased wave attack and the onset of waterlogging and anaerobic conditions on mature marsh. Some support for the latter view comes from the fact that in areas where the introduction has

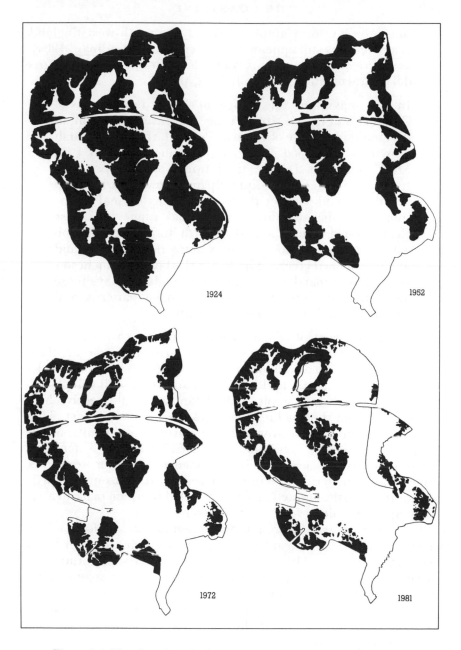

Figure 8.6 The changing distribution of *Spartina* dominated marsh in Holes Bay, Poole Harbour, Dorset, at four dates (after Gray and Pearson, 1984, figure 1).

been more recent, for example in Wales and north-west England, the area of *Spartina* still appears to be increasing (Deadman, 1984).

The nature of salt marshes is controlled to a considerable degree by tidal characteristics (Ranwell, 1972, p. 28):

> In the maximum tide range of 12 m in the Bristol Channel, salt marsh growth ranges vertically over 4 m; in the small tidal range of 1.8 m in Poole Harbour salt marsh growth is telescoped into a vertical range of about 1 m. Marshes within a large tidal range tend to be more steeply sloping and consequently have more clearly zoned vegetation and sharper drainage systems normal to the shore. Marshes within a small tidal range have less clearly zoned vegetation and sluggish drainage on the ebb which produces a more complex network of winding and much branched creeks. The Poole Harbour to Chichester Harbour salt marshes in a minimum tidal zone have those distinctive features of complex drainage patterns and mosaics of indistinctly zoned vegetation.

Another major control of marsh type is position with respect to wave action and inputs of fresh water. *Maritime salt marshes* are those which develop on relatively open coast conditions. Those in the lee of small islands offshore and found in association with shingle spit and dune growth (e.g. Scolt Head Island, Norfolk) receive little fresh water from the low hills that back the mainland coast. They are also developed on relatively coarse particled sediments, as one might expect in the conditions of relatively strong water flow. These marshes, because of their dependence on shelter from spits, dunes and shingle ridges, often have restricted horizontal growth, so that their vertical growth is not paralleled by an increase in area. They are thus called *closed marshes* (Steers, 1977).

Estuarine mouth marshes are a type of open marsh and form within estuaries in the lee of coastal spits (e.g. those on the Dovey or Burry estuaries in Wales). They tend to have a coarser substratum than those further up estuary and are exposed to a less strong saline influence than the maritime type.

Mid-estuary marshes are usually developed on silt and are subject to advance or retreat under the influence of changes in the low water estuary channel.

Upper estuarine marshes are most sheltered of all, have greatly reduced salinity and develop on clay-silt substrata (Ranwell, 1972, p. 33).

Embayed marshes can be extensive in area when they occur on a shallowly embayed coast with relatively freely draining silt and high salinity (e.g. Morecambe Bay). By contrast, narrow-mouthed deep embayments, being more sheltered, allow deposition of finer particles, producing marshes with intricate drainage systems (e.g. Chichester Harbour, Poole Harbour, or Hamford Water in Essex).

Finally, it is worth noting that a contrast has often been drawn between west-coast and east-coast marshes (e.g. Marshall, 1962; Steers, 1960). The former are generally believed to be made of sandier sediments and to have a more grassy sward; the latter are thought to be made of rather sloppier mud and have steep-sided creeks and a more diverse vegetation cover. The utility of this generalization may be revealed by further investigation. Certainly, west-coast marshes, with extensive development of communities with *Puccinella maritima* and *Juncetum gerardii*, are much grazed, causing grass to predominate (Adam, 1978, 1981), although grazing also occurs elsewhere, notably in Essex (Williams and Hall, 1987) and in the vicinity of the Wash. Equally, material properties appear to control the form of salt marsh cliffs (Allen, 1989), so that the cliffs of the muddy Severn Estuary marshes are mostly strong and tall because of their cohesiveness, whereas in the mainly sandy Solway Firth and Morecambe Bay systems the marsh cliffs are only strong in their upper parts, where a dense root-mat of marsh grasses binds the sediments. For the most part they are weak and low, and are degraded by cantilever and toppling failures.

Shingle spits

The British coast displays many excellent examples of spits (see table 8.4) and related constructional forms. They are not, however, ubiquitous (figure 8.7). There is a clear process control of their distribution. Pethick (1984) proposes that in coastal areas which experience tidal ranges of over 4 m, tidal landforms, which include tidal flats and saltmarshes, will be dominant. On the other hand, if the tidal range is less than about 2 m then wind waves provide the dominant coastal process, so that beaches, spits and barrier islands will be the dominant coastal landforms. The tidal range map (figure 8.8) indicates that only limited areas experience a tidal range of less than 2 m: the east Norfolk coast, the south coast from Start Point to the Isle of Wight, and part of the Welsh coast.

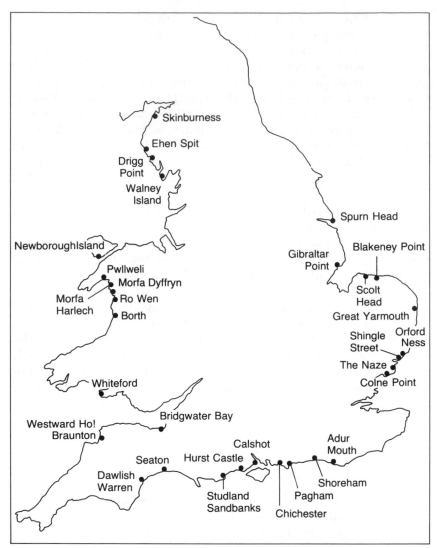

Figure 8.7 The distribution of major sand and shingle spits in England and Wales.

In general terms the distribution of spits broadly coincides with these areas.

Whether a spit or some other form of shingle accumulation occurs also depends on the discharge that comes down the estuary

Table 8.4 Select examples of spits in England and Wales

Spit	Reference
Stert Point, Bridgwater Bay	Kidson (1960)
Dawlish Warren, Devon	Kidson (1950)
Blakeney Point, Norfolk	Hardy (1964)
South Haven Peninsula and Sandbanks	Robinson (1955)
Pagham and Chichester	Robinson (1955)
Gibraltar Point, Lincs	King (1978)
Scolt Head Island, Norfolk	Steers (1960)
Spurn Head	De Boer (1964)
Hurst Castle	King and McCullagh (1971)
Shingle Street, Suffolk	Randall (1973)
Orford Ness	Carr (1965)

across which sediment movement takes place. Where the discharge from the estuary is negligible, the transported sediment succeeds in closing the gap across the coastal inlet, producing a shingle beach with a lagoon behind (e.g. Loe Pool in Cornwall and Slapton Ley in Devon). At the other extreme, in great estuaries like the Humber and the Mersey, there is no possibility that a spit or bar could close the inlet. In between are examples where the opposing forces are so nicely balanced that one spit may be breached periodically by either the sea or the river which it has succeeded in diverting (Kidson, 1963).

A shingle structure which is rather different from the spits described thus far is Dungeness in south-east England. This is the largest cuspate shingle foreland in Britain. It appears to have originated when the Flandrian transgression flooded an embayment formed on a platform of Hastings Beds and Weald Clay. A cliff-line was formed, now degraded, in a great arc running from Winchelsea, behind Rye and Appledore to Hythe. The north-east growth of a spit across the embayment, to the vicinity of Hythe, led to accretion in a marsh and swamp environment. The shingle structure gradually evolved through the centuries to give the present complex structure (Green, 1968). Evolution has been rapid since Roman times (Cunliffe, 1980) as figure 8.9 shows. It is a feature of considerable ecological importance but it has been drastically maltreated and degraded by human activities in recent decades and centuries (Ferry and Waters, 1985).

Another major structure formed by the Flandrian transgression is Chesil Beach in Dorset. This is not a spit, since it possesses no laterals, and appears to have evolved as a barrier beach which moved onshore as the transgression progressed, with the Fleet Lagoon developing between it and the mainland. Much of the sediment will have been combed up from the sea floor as the sea-level rose and may not, therefore, be receiving much replenishment given the currently relatively stable state of sea-level (Carr and Blackley, 1973). Many of the shingle features of England and Wales may also be essentially relict features, so that their life may be endangered if excessive sediment removal occurs (see p. 323).

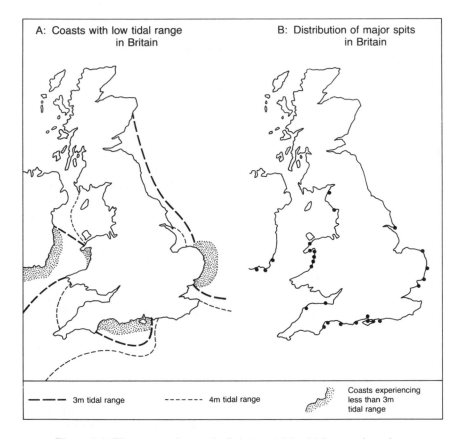

Figure 8.8 The areas of coast in Britain with a tidal range less than 3 m (A) are also areas noted for their spit development (B) (after Pethick, 1984, figure 4.17).

Plate 27. There are a number of shingle structures, largely composed of flint, on the coast of East Anglia. The growth of Orford Ness has caused the River Alde to be deflected southwards. (Cambridge University Collection: copyright reserved)

That there are so many shingle structures and gravel beaches in England and Wales may well also be a function of this Pleistocene inheritance, the coarse sediment being produced initially by glacial and periglacial processes (Orford, 1987).

Coastal dunes

The distribution of coastal dunes shown in figure 8.10 has been mapped from the 1:650,000 scale Quaternary map of Britain, a

Plate 28. The *Flandrian* transgression of the Holocene produced a series of shingle accumulations along the English coast. This example, which enclosed Slapton Ley, is at Torcross in Devon. (Tony Waltham)

review of the literature (see table 8.5) and the study of 1:50,000 OS topographic sheets. Ranwell and Boar (1986) estimate that dunes constitute around 9 per cent of the total coastline of England and Wales, extending over 400 km of coast length.

THE DUNES OF THE CHANNEL COAST OF ENGLAND

Dunes have a relatively limited distribution along the Channel coast of England between Land's End and the Thames Estuary. The reasons for this are not entirely clear, but large stretches of the coastline are backed by cliffs (and so do not have suitable conditions for dune accumulation); flinty shingle structures are relatively common; neatly compartmentalized bays of the type found in Cornwall and Pembrokeshire are less well developed for dunes to form behind; the tidal range is lower and thus exposes a smaller intertidal zone for deflation, and inputs of glacial and fluvioglacial debris for subsequent reworking are less likely because of the area's position with respect to the great Pleistocene ice-sheets. Likewise wave erosion of the long stretches of Chalks and other limestones may produce very little quartzose material of sand size, and many beaches are formed of large quantities of flints derived from the erosion of the Chalk and of Tertiary beds. These tend not to form dune-building materials.

The most important feature of this group of dunes is that they are low in carbonate content. This is especially true of the dunes at Studland (South Haven Peninsula) which have a carbonate content that is probably the lowest found on the British coastline (< 1 per cent). This is reflected in the very marked development of an acid-tolerating flora, including *Calluna*. The explanation for this is that the Studland dunes have developed across a bay, Poole Harbour, which is cut into the highly quartzitic, highly acidic and highly erodible sands of Tertiary Beds, which underlie the Dorset heathlands. Likewise the Littlehampton and Sandwich dunes are developed in areas of Tertiary Beds, and the Camber Sands are developed in an area of Wealden (Cretaceous) sediments.

THE NORTHERN COAST OF THE SOUTH-WEST PENINSULA

The northern coast of the south-west peninsula between Land's End and Ilfracombe has a large number of dune fields, most of

Table 8.5 Selected studies of dune areas in England and Wales

Reference	Location
Bhadresa (1977)	Holkham, Norfolk
Brown et al. (1985)	Talacre Warren, Clwyd
Diver (1933), Wilson (1960)	South Haven Peninsula, Dorset
Firth (1971), Pizzey (1975)	Camber Sands, Sussex
Gorham (1958)	Blakeney, Norfolk
Greenwood (1972, 1978), Kidson and Carr (1960)	Braunton Burrows, Devon
Gresswell (1937)	SW Lancashire
Hepburn (1944)	Camel Estuary, Cornwall
Higgins (1933), Stuart (1924)	S Wales (general)
Kennard and Warren (1903)	Newquay, Cornwall
Lambert and Davies (1940)	Dovey Estuary, N Wales
Lees (1982), Potts (1968)	Gower, S Wales
Moore (1931)	Isle of Man
Pearsall (1934)	N Lancashire
Permadasa et al. (1974)	Aberffraw, Anglesey
Randall (1961)	Merthyr Mawr, S Wales
Ranwell (1959)	Newborough Warren, Anglesey
Roy (1967)	Scolt Head Island

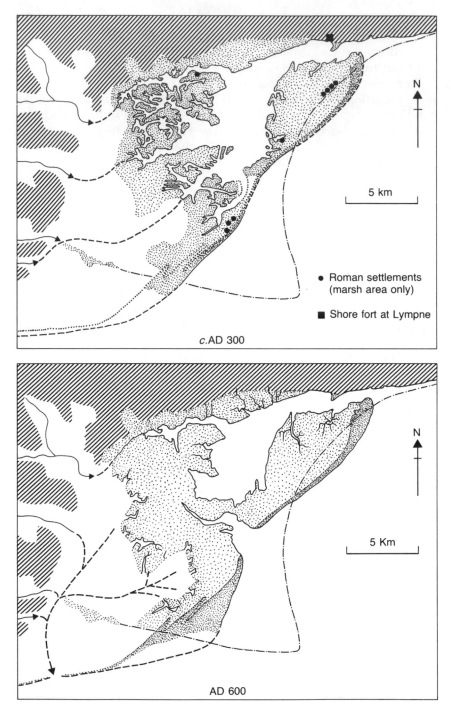

N

5 km

● Roman settlements
  (marsh area only)

■ Shore fort at Lympne

*c.*AD 300

N

5 Km

AD 600

Figure 8.9 Stages in the evolution of the great cuspate foreland of
Dungeness (the Romney Marsh region) since Roman times (after
Cunliffe, 1980, figures 17, 19, 20 and 22).

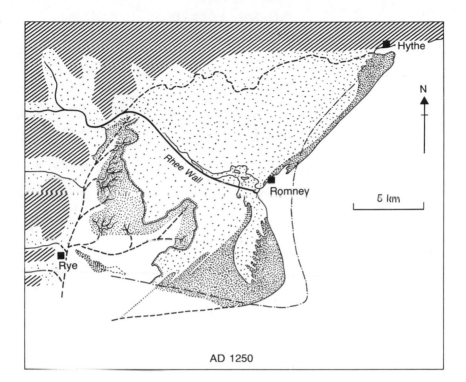

AD 1250

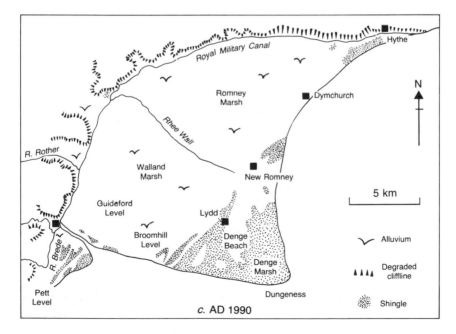

c. AD 1990

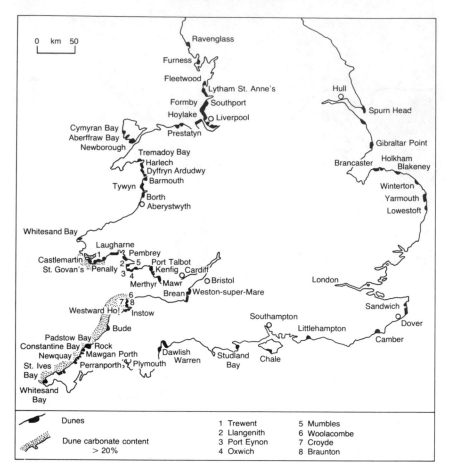

Figure 8.10 The distribution of coastal sand dunes in England and Wales.

which occur at the head of bays, but some of which occur in estuaries (e.g. Rock and Instow) and others on prograding shingle structures (e.g. Northam Burrows at Westward Ho!). This group of dunes is distinctive because of the generally high carbonate content of the dune sands (table 8.6). The maximum carbonate content (87.5 per cent) occurs at Constantine Bay but there are other locations where the carbonate content of the dunes is over 50 per cent (St Ives Bay, Perranporth, Newquay, Mawgan Porth, Rock and Bude). The carbonate content decreases on the northern part

Table 8.6 Sand characteristics from coastal dunes in England and Wales. The $CaCO_3$ values are determined from the content soluble in HCl (10%), and the D_{50} values are the median grain size in ϕ units. All samples come from active crests of foredunes

Location	$CaCO_3$ (%)	D_{50} (ϕ)
(a) Northern coast of the SW Peninsula		
Whitesand Bay	30.96	1.19
St Ives Bay (Hayle)	56.80	1.56
Perranporth	52.88	1.93
Newquay	76.85	0.83
Mawgan Porth	61.04	1.81
Constantine Bay	87.50	1.20
Rock	72.62	2.11
Bude	63.74	1.56
Westward Ho! (Northam)	21.79	2.45
Instow	19.52	2.33
Braunton	19.59	2.13
Croyde	12.81	1.87
Woolacombe	21.96	1.82
Mean ($n = 13$)	46.00	1.75
(b) Bristol Channel coastline		
Brean	12.82	2.52
Weston-super-Mare	11.65	2.62
Merthyr Mawr	8.49	2.41
Kenfig	9.86	2.15
Oxwich	12.45	1.93
Llangennith	15.65	1.63
Pembrey	12.04	2.33
Pendine (Laugharne)	11.15	2.40
Mean ($n = 8$)	11.76	2.25
(c) Pembrokeshire		
Pennally (Tenby)	9.65	2.19
Trewent	27.39	1.82
Broadhaven (St Govan's)	29.87	1.79
Freshwater West	40.98	2.02
Mean ($n = 4$)	26.97	1.96

Location	$CaCo_3$ (%)	D_{50} (ϕ)
(d) Welsh coast north of Aberystwyth		
Borth	4.98	2.29
Twwyn	5.72	1.98
Barmouth	4.78	2.09
Duffryn Ardudley	3.34	2.31
Harlech	3.96	2.13
Tremaoog (Morfa Bychan)	4.35	2.64
Newborough	4.56	2.50
Aberffraw	13.20	2.47
Cymyran	5.12	1.95
Mean ($n = 9$)	5.56	2.26
(e) Coast of North Wales and NW England		
Prestatyn	5.36	1.88
Bridge of Ayr	6.67	1.93
Formby	5.00	2.30
Ainsdale	3.57	2.13
Birkdale	5.84	2.13
Lytham	3.87	2.18
Barrow	1.51	2.21
Ravenglass	1.08	2.24
Mean ($n = 8$)	4.11	2.13

of this coastline, declining to around 20 per cent as the Bristol Channel is approached.

The high carbonate content of the dunes seems to have been long established for it contributes to the formation of quite extensive spreads of Pleistocene dune rocks (aeolianites), notably at Saunton Downs and near Croyde. The carbonate-rich nature of this group of dunes is related to the elevated carbonate content of the beach sands. The explanation for this cannot be sought in terms of inputs from carbonate-rich bedrocks, for these are rare along this area of coastline. Rather the area offshore is extremely rich in shell debris and forms part of the high-carbonate environment noted by Merefield (1984). In this high energy wave-dominated environment carbonate skeletons are broken up and then driven onshore, making them available for incorporation in upper beaches and dunes (Giles and Pilkey, 1965). Furthermore, along portions of the coast there is a variable supply of sand-sized terrigenous material, depending largely on the relative durability and weathering potential of local lithologies. This dictates the scale of carbonate dilution by terrigenous material. Between the Camel Estuary and Land's End – the zone of maximum carbonate concentration – sandstones are rare; the coastline is dominated by granites, which are relatively resistant, and by slates and shales, which on breakdown cause the production of very limited quantities of durable sand-sized material. The dune sands of this stretch of coastline are also notable for their relative coarseness, with the mean D_{50} value being 1.75 ϕ. This in turn seems to be related to the presence of large quantities of carbonate grains which, on account of their relatively low density and sometimes platy nature, are transported with some ease by the wind. However, those examples which are in estuarine positions (e.g. Westwood Ho! and Braunton at the mouth of the Taw-Torridge system; Rock, within the Camel Estuary; and Instow within the Torridge estuary) are relatively fine-grained (D_{50} = 2.11–2.45 ϕ), reflecting their partial source in estuarine alluvia.

THE DUNES OF THE BRISTOL CHANNEL COASTS

On account of their high tidal range, the Bristol Channel and the mouth of the River Severn, display broad intertidal zones. On the south side of the Channel there is a group of dunes between Burnham-on-Sea and Weston-super-Mare, and these lie to the lee of the Berrow Flats. Another major group, which includes Merthyr Mawr, Kenfig Burrows and Margam Burrows, occurs on the south-west-facing coast between Ogmore-by-Sea and Port Talbot. In addition, important dunes occur behind some of the major bays of the Gower Peninsula (e.g. Oxwich and Llangennith), and to the lee of the broad intertidal flats at the mouths of the Loughhor (Llwchyr) and Tywi estuaries (e.g. Pembrey and Pendine sands).

The dunes of this area (see table 8.6b) are remarkably similar in their carbonate content, which averages 11.76 per cent. Thus they are relatively depleted in carbonate compared with those of the north coast of the south-west peninsula. On the other hand they are richer in carbonate than the dunes of the Welsh coast to the north of Aberystwyth, of the Lancashire coast, or the coasts of southern and eastern England. One probable reason for this intermediate carbonate level is that the rocks of this portion of the coastline include large outcrops of Carboniferous limestone.

The grain size characteristics of the dunes indicates that the sands are somewhat finer than those of the south-west peninsula, with those at the mouth of the Severn, at Brean and Weston-super-Mare, being notably finer than the average ($D_{50} = 2.52 \phi$). This reflects the inputs of relatively fine-grained material in the estuarine situation.

THE DUNES OF PEMBROKESHIRE

The indented coastline of Pembrokeshire possesses a series of dune systems, with the most impressive examples occurring behind the broad, south-west facing expanse of Freshwater West Bay.

Three of the four examples given in table 8.6c show significantly elevated carbonate contents in comparison with those of the British Channel coastline, and two of them (Broadhaven and Freshwater West) have comparable values to some of those found along the south-west peninsula in Cornwall. The coastline is one of relatively resistant lithologies, so that dilution by non-carbonate sands is relatively low, and there are some outcrops of Carboniferous Limestone in the immediate vicinity of Broadhaven.

The grain size characteristics of the dune sand (mean $D_{50} = 1.96 \phi$) reflect this relatively high carbonate content by being coarser than the norm.

THE WELSH COAST NORTH OF ABERYSTWYTH

Many of the most imposing dune fields on the Welsh coast north of Aberystwyth, including those of Anglesey, have developed in association with bar and spit growth across the mouths of major estuaries (e.g. Borth Spit, the Bar near Barmouth, Morfa Dyffryn, Morfa Harlech).

The carbonate contents of the dunes (table 8.6) are, with the exception of Aberffraw Bay in Anglesey, generally low, averaging only 5.56 per cent. The Aberffraw value may possibly be accounted for by its isolation in a narrow bay, partially cut off from major inputs of fluvial sands. There is also a small outcrop of Carboniferous Limestone in the vicinity. With this exception, most of the dunes and beaches have clearly been affected by inputs of glacial and fluvioglacial deposits, together with modern fluvial inputs from the major river systems draining the mountainous, largely non-calcareous areas of North Wales. The sand from three sites – Tremadog, Newborough and Aberffraw – is notably fine grained.

The dunes of North Wales and north-west England

Some of the most impressive dune areas in the British Isles occur between the Dee and Lune estuaries in north-west England. It is perhaps not coincidental that their occurrence parallels the outcrop of the longest stretch of Permo-Triassic marls and sandstones in Britain. These beds would provide a ready source of beach sand with suitable grain-size characteristics to promote dune formation either through direct cliff erosion or by fluvial inputs from sandstone catchments. Furthermore, being dominantly quartzose, they could be an important factor in explaining the low carbonate contents of the dunes (mean = 4.11 per cent; table 8.6).

The main types of dune system

The dune systems discussed in the foregoing regional analysis can conveniently be classified into five main types, following the scheme of Ranwell and Boar (1986). These are illustrated in figure 8.11.

Offshore island dune systems are those developed on offshore or barrier islands; they serve to protect mudflats which lie in their lee. They are narrow, subject to overwash from time to time and may form an age series extending in one direction along the coast as at Blakeney, Norfolk.

Prograding ness dunes form from an open coast where there is an abundant supply of sand at an accumulation point (ness) receiving sand by longshore drift from two directions at once. These conditions pertain on parts of the eastern coast (e.g. at Winterton Ness in Norfolk), where the prevailing wind blows offshore and is in opposition to the dominant wind.

Spit dune systems form on sandy promontories at the mouths of estuaries, and often form a fan-like series of dune ridges and intervening slacks, with the 'handle' of the fan tied to the mainland. Examples include Whiteford Burrows, Glamorgan, and the Studland Dunes at South Haven Peninsula in Dorset.

Bay dunes, the commonest type, accumulate in bays developed along indented coastlines like those of the south-west peninsula and Pembrokeshire.

Finally, hindshore dune systems are found on extensive sandy coasts where the prevailing wind is also the dominant one. Large dunes are driven inland as great arcs or ridges, such as those of Braunton Burrows in Devon and Newborough Warren in Anglesey.

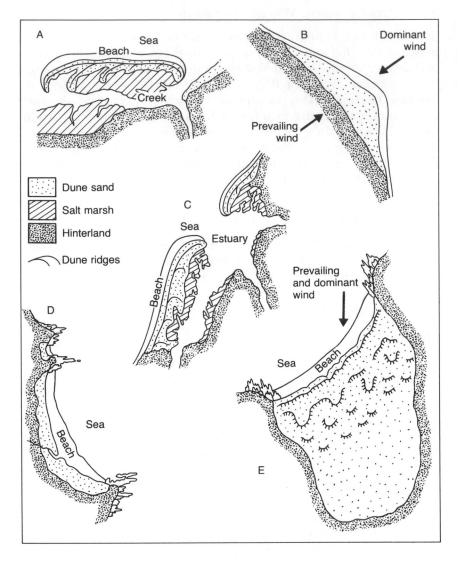

Figure 8.11 Different types of sand dune in England and Wales (after Ranwell and Boar, 1986, figure 4): (A) offshore island dunes (e.g. Scolt Head Island, Norfolk); (B) prograding ness dunes (e.g. Winterton Ness, Norfolk); (C) spit dunes (e.g. South Haven peninsula, Dorset; Whiteford Burrows, Glamorgan); (D) bay dunes (e.g. Oxwich Bay, Gower); (E) hindshore dunes (e.g. Braunton Burrows, Devon; Newborough Warren, Anglesey).

Shore platforms

About 20 per cent of the coastlines of England and Wales are fringed by gently sloping rock cut surfaces called shore platforms (figure 8.12; Trenhaile, 1974). Although such features are sometimes termed 'wave-cut platforms' this is an essentially misleading terminology, for in addition to wave attack, there are many other processes that contribute to the planation, including miscellaneous

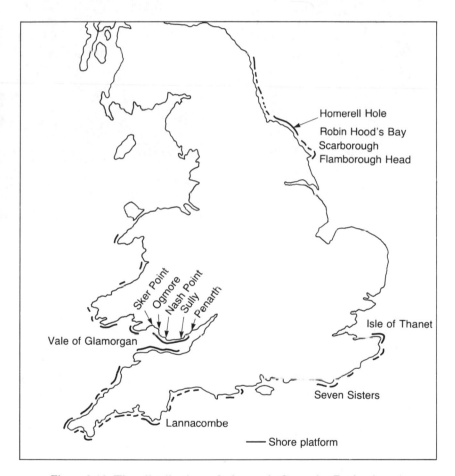

Figure 8.12 The distribution of shore platforms in England and Wales (from *The Atlas of Britain and Northern Ireland*, Clarendon Press, 1963, p. 22).

Plate 29. Robin Hood's Bay, on the Yorkshire coast, is an unstable piece of coastline developed in incompetent sedimentary rocks with a capping of Pleistocene drift. It displays a magnificent shore platform. (Michael J. Stead)

weathering processes. For example, Robinson (1977a) suggested that a contributory factor to the formation of shore platforms developed in Lias shales in north-east Yorkshire was desiccation and contraction during intertidal periods and hydration and expansion during flood tide periods. These forces cause cracking of the shale bedding laminae so that the fragments thereby produced can be removed by waves. On schist platforms in Devon, at Prawle, salt-weathering may contribute to the planation process (Mottershead, 1982), while chalk shore platforms in south-east England are rapidly modified in harsh winters by combined salt and frost attack (Robinson and Jerwood, 1987).

The distribution of the platforms is by definition restricted to rocky coasts, so that they are absent from accretionary coasts and coasts formed in unconsolidated drift. The most important areas are the coast of north-east England to the north of Flamborough Head; the Chalk coasts of south-east England; the exposed coast of the south west peninsula; and the coastline of the Bristol Channel (see figure 8.13).

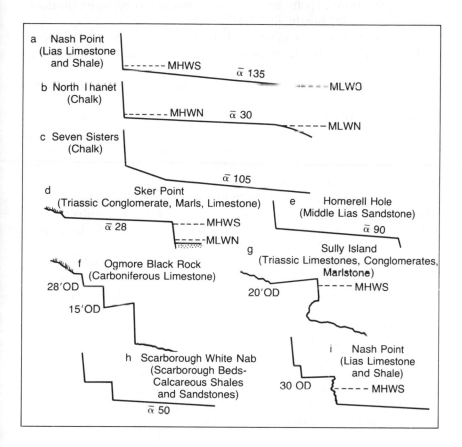

Figure 8.13 Shore platform and ledge profiles in England and Wales. $\bar{\alpha}$ is the mean platform gradient in minutes. Locations are shown in figure 8.12 (after Trenhaile, 1974b, figure 2).

A strong, positive relationship has been identified between tidal range and platform gradient. Using more than 800 surveyed profiles from three major macrotidal areas (Vale of Glamorgan, south-east England, and the north-east coast between Flamborough Head and Whitburn) Trenhaile (1974a) found a correlation of +0.92

between local means of platform gradient and tidal range. Another control of gradient may be the length of wave fetch perpendicular to coast orientation. A possible explanation for this inverse correlation is that greater fetch produces greater energy, which produces more effective planation and therefore lower gradients. A further correlation coefficient of + 0.79 was noted between platform gradient and cliff height for chalk areas.

The situation is less clear with respect to controls of platform width. There is little agreement as to whether shore platforms in England and Wales get wider as tidal range increases. In southern England the widest platforms occur where wave action is most vigorous. The same is true of the Vale of Glamorgan in South Wales where platforms facing the dominant and prevailing south-westerly winds are between 150 and 400 m in width, whereas those of sheltered, easterly facing coasts are much narrower (Trenhaile, 1987). The platform of north-east Yorkshire, developed in Lias beds, attains widths of 300 m (Robinson, 1977b).

Platform slopes are also affected by topographic situation. Chalk platforms on the Isle of Thanet are steeper in the embayments, presumably because wave action is weaker. However, rock resistance is also a factor, with more resistant rocks tending to give steeper platforms, so that in the Vale of Glamorgan platforms are steeper on the headlands, probably because differences in rock resistance are in that particular case more significant than differences in wave intensity.

Lithology and structure are other important controls of platform morphology. Platforms will be ill developed either on highly resistant rocks or on rocks of very limited strength. The best platforms occur on rocks of moderate resistance which are subjected to a fairly vigorous wave environment. Steps and slopes on the platforms will be greatly influenced by the dip of beds, and their relative thicknesses and resistances.

Sea-level changes

The effects of sea-level change can be seen all round the coastline of Britain. There are various submerged coastal features such as the drowned river mouths (rias) of Cornwall and Pembrokeshire; the buried channels at the mouths of many rivers; notches and benches in submarine topography (Donovan and Stride, 1975); and remnants of forests or peat layers at or below present sea-level. Elsewhere evidence of emerged coastal features is provided by high-level shore platforms and stranded beach deposits. Many stretches of coast show the evidence of both emergent and submergent phases in their history.

Plate 30. The raised beach at Portland Bill, Dorset. (Author)

The causes of such sea-level changes may be either dominantly world-wide or more local. The world-wide (eustatic) causes include those oscillations that took place as a response to the quantity of water stored in ice-caps during glaciations and deglaciations (glacio-eustasy) of the Pleistocene. On a longer time-scale, sea-levels world-wide may have changed in response to changes in the volumes of the world's oceans caused by plate tectonic movements (orogenic eustasy).

During glacials, when the volume of ice caps was approximately three times the present figure, sea level would have been lowered substantially. Donn et al. (1962) estimate that in the penultimate glaciation sea-levels might have been 137 to 159 m below current sea-level, and that in the last glaciation (the Devensian) the maximum lowering would have been between 105 and 123 m. During interglacials, sea-levels would tend to be at broadly similar levels to those of the present Holocene interglacial, although if previous interglacials were warmer than today sea-levels might have been rather higher.

The role of orogenic eustasy is difficult to quantify, but there is some evidence to suggest that it may have caused a decline in global sea-levels during the Pleistocene. This mechanism is a possible explanation for the presence of early Pleistocene high-level platforms at heights up to 180 m that have been postulated by Wooldridge and Linton (1955) in south-east England and by Brown (1960b) in Wales. On the basis of recent studies of global tectonics, Bloom (1971, p. 355) estimated that 'the spreading of the ocean basins since the last Interglacial could accommodate about 6 per cent of the returned meltwater, and the post-glacial shorelines would be almost 8 m lower than the interglacial shorelines of 100,000 years ago.'

The general eustatic picture of sea-level change is complicated by local tectonic activity. Mitchell (1977) has identified at least three district provinces in the British Isles. The first province is the North Sea basin, where there has been tectonic sinking. This still continues and has allowed large thicknesses of sediment to accumulate. The second province is the 'stable Celtic Sea area', along whose drowned coastline it is possible to trace a remarkably level wave-cut shore platform, which must be of considerable antiquity. The third province Mitchell identifies is the northern part of Great Britain and Ireland, where the growth and decay of Pleistocene ice caps caused isostatic sinking and rebound.

However, the concept of a 'stable Celtic sea area' is not one that receives universal support. Tooley (1978, pp. 177–8) points out that tide gauge data indicate that subsidence at Newlyn, Cornwall, is currently proceeding at a rate of over 1 mm per year and highlights Walcott's (1972) contention that the effects of ice loading will be recorded up to 1,500 km beyond the ice limits of the last glaciation. If Walcott is right, the south-western parts of Britain might well suffer post-glacial subsidence as a compensation for the isostatic updoming taking place in Scotland and northern England.

Much more widely acceptable is the concept of a subsiding North Sea Basin. There are up to 1,000 m of Pleistocene sediments in the southern part of the basin (Lovell, 1986, p. 184; see figure 8.14) and associated warping may have affected south-east England (Jones, 1981, pp. 146–7). Analyses of sediments in the North Sea Basin itself and in the Fens indicates that sinking is occurring at rates of up to 0.9 m per 1,000 years: Shennan (1986) has calculated 0.9 m^{-1} y^{-1} (Holocene) for the Fenland; Sejrup et

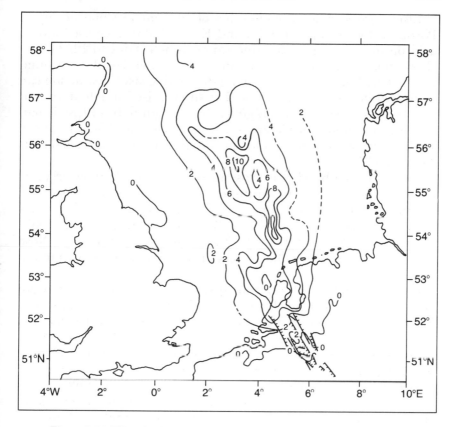

Figure 8.14 The thickness of Quaternary sediments in the North Sea basin. The contour interval is 100 m. Compare with figure 1.11 (after Lovell, 1986, figure 8.8).

al. have calculated 0.6–0.9 $m^{-1} y^{-1}$ (Pleistocene) for the Central North Sea Basin.

The age of the widespread raised beaches of Wales and south west England has been the subject of some controversy, not least because precise radiometric dating of ancient beach deposits has proved impossible. Bowen (1973) favours a relatively simple sequence of events involving (1) glaciation, (2) interglacial raised beach formation, and (3) cold climate head formation and redistribution of old glacial deposits by solifluction. He argues that there are no grounds for consigning the head to more than an old cold period; that the head is of Devensian age; and that logically the raised beach dates to the previous (Ipswichian) interglacial.

Elsewhere, several raised beaches of different ages may occur. At Portland, for example, Davies and Keen (1985) have used amino acid dating to suggest that one raised beach dates to 125,000 BP (Ipswichian) and another to an earlier interglacial at around 210,000 BP. Likewise, further along the south coast a distinction can be drawn between the lower or 'Brighton' raised beach at around 8 m above sea-level, which is generally thought to be of Ipswichian age, and the upper or Goodwood raised beach at 23–38 m above sea-level, which may date to the penultimate interglacial (Shephard-Thorn and Wymer, 1977).

Even if raised beaches in close proximity to each other occur at approximately the same altitude, there is no good reason for assigning them all to the same age. Preliminary amino acid studies from the raised beaches of Tor Bay (Mottershead et al., 1987) demonstrate this clearly, with the classic site at Hope's Nose (Torquay) being older than others at the same altitude within the area.

One consequence of low-glacial sea-levels was that rivers were able to cut down their channels deeply below present sea-levels. Flandrian sedimentation may then have partially filled the channels with alluvial and estuarine materials, creating buried channels. Such features are especially well known in the Bristol Channel area (G. J. Williams, 1968) and some data on the maximum depths of buried channels at present river mouths are given in table 8.7.

The rapid rise of sea-level in the Holocene caused the reworking of sediments deposited on the present continental shelf by the rivers of glacial times. These sediments were combed up as the transgression progressed to create some major sedimentary features such as Chesil Beach, Dungeness, Orford Ness and Dawlish Warren. In addition, the sea flooded lowlying areas, including the lower reaches of river valleys, to create rias (especially in southwest England and Pembrokeshire) and embayments (e.g. the present sites of Romney Marsh, the Pevensey Levels, the Broads, the Fens and the Somerset Levels). These have subsequently become filled with alluvial sediments or cut off from the sea by the growth of spits (e.g. Poole Harbour). The rising sea flooded the floor of the North Sea and the English Channel (c.9,600 BP), achieving the final separation from the Continent in about 8,600 BP (Jones, 1981) and the almost total inundation of the Thames Estuary by 8,000 BP. The lateral rate of transgression across the North Sea was probably around 60 km 1000 y^{-1}, while in the Bristol Channel it was probably 15 km 1000 y^{-1}.

Table 8.7 Depths of buried channels in
the Bristol Channel area and south-east
England

Location	Depth (m OD)
Usk (Newport)	> − 19.8
Rhymney (Cardiff)	− 8.5
Taff (Cardiff)	− 12.8
Ely (Cardiff)	− 11.6
Wye (Chepstow)	− 17
Tawe (Swansea)	− 61
Bristol Avon	− 19.8
Taw/Torridge	− 30.5
Severn (Severn Tunnel)	− 22.2
Thames	− 47
Sussex Ouse (Newhaven)	− 29.5
Arun (Arundel)	− 33.5
Exe (Devon)	− 52

Sources: Various in Williams (1968), Durrance
(1969), Jones (1981)

In very general terms the temporal pattern of Holocene
(Flandrian) transgression is comparable in different areas of
England and Wales. Figure 8.15 illustrates how in the last 10,000
years there has been a phase of very rapid sea-level rise to around
6,500 years BP, followed by a rather gentle rise, or even stability,
thereafter. However, in detail the pattern is more complex (as
revealed by the stratigraphy of areas like the Fens) and there are
also differences between areas, caused by subsidence (in the south
east) and by isostatic uplift (in the north). At certain times the rate
of sea-level rise must have been very rapid. Tooley (1978, p. 200)
suggests that during his Lytham III transgression in north-west
England (c.7,600–7,200 BP) sea level may have risen by over 7 m in
just 200 years, possibly as a consequence of the break-up of the
Laurentide Ice Sheet over Hudson Bay.

The sedimentary record in lowland areas is frequently complex.
This is exemplified in the Fens (Shennan, 1986) where there have
been multiple sequences representing either positive tendencies
of sea-level (described as WASH) or negative (described as
FENLAND). These are shown in table 8.8.

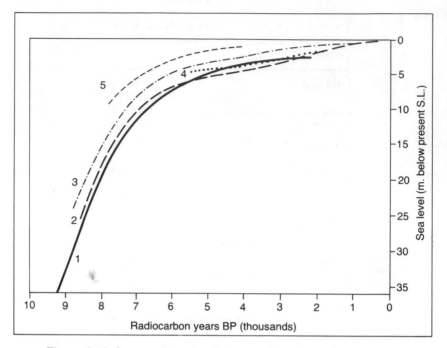

Figure 8.15 Curves of sea level change since 10,000 years BP for
different parts of England and Wales: 1 = Bristol Channel; 2 =
English Channel; 3 = Cardigan Bay; 4 = Somerset Levels; 5 = North
Wales. The datum level is MHWST (mean high water spring tide),
(after Shennan, 1983, figure 8).

Table 8.8 Holocene sea-level tendencies in the East
Anglian Fens

Period	Tendency
> 6,300 BP	WASH I
6,300–6,200	FENLAND I
6,200–5,600	WASH II
5,600–5,400	FENLAND II
5,400–4,500	WASH III/IV
4,500–4,200 (3,900)	FENLAND IV
4,200 (3,900)–3,300	WASH V
3,300–3,000	FENLAND V
3,000–1,900	WASH VI
1,900–1,550	FENLAND VI
1,550–1,150	WASH VII
1,150–1,000(950)	FENLAND VII
950 onwards	WASH VIII

Source: data in Shennan (1986)

The sediment in the Broadland valleys of eastern Norfolk preserves evidence of the main phases of Holocene marine transgression, represented by two layers of estuarine, clay sandwiched between layers of freshwater peats (Coles and Funnell, 1981). The lower clay is 12–16 m thick at its seaward limit, extends 20 km inland, at its upper surface (dated at *c*.4,500 BP) and is at present at −5.5 to −6.5 m below ordnance datum. The upper clay, which is 6 m thick

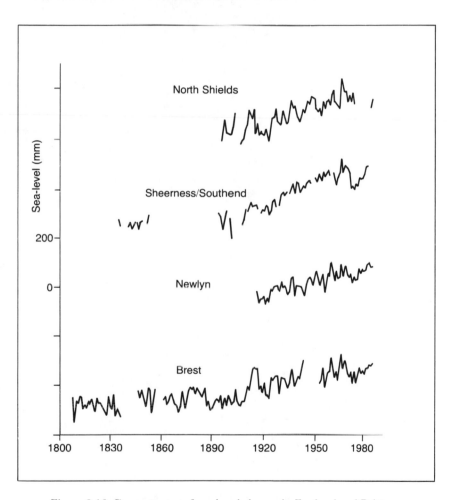

Figure 8.16 Current rates of sea-level change in England and Brittany (France) (modified after data from Proudman Oceanographic Laboratory in the *Corporate Plan* of the Natural Environmental Research Council, 1989, p. 14).

at the coast and extends 23 km inland, dates to around 2,000 BP, while the upper surface dates to about 1,500 BP.

The study of the tide-gauge record suggests that sea-level continues to rise in England and Wales (figure 8.16). The trend for Newlyn in Cornwall is around 17 cm per century (Natural Environment Research Council, *Corporate Plan*, 1989, p. 14). If greenhouse warming takes place, this rate may increase fourfold in the next century, with substantial consequences for evolution of the coastline.

9

THE HUMAN IMPACT

Introduction

The human role in creating landforms and modifying the operation of geomorphological processes is a matter of increasing importance as population increases, as new technology is introduced and as energy is harnessed to an ever-increasing degree. The range of the human impact on both forms and process is considerable. Table 9.1 lists some anthropogenic landforms together with some of the causes of their creation. There are indeed very few spheres of human activity which do not, even indirectly, create landforms. However, it is useful to recognize that some features are directly produced by anthropogenic processes. These tend to be relatively obvious in their form and origin and are frequently created

Table 9.1 Some anthropogenic landforms

Process	Form
Mining, marling, dew ponds	Pits and ponds
Peat extraction	Norfolk Broads
Mining	Spoil heaps, subsidence depressions
Ploughing	Terracing, lynchets, ridge-and-furrow
Transport	Cuttings, embankments, canals
River and coast management	Dikes, embankments
Defence	Mounds, craters, moats
Water management	Reservoirs

Plate 31. Penrhyn Slate Quarries, North Wales, from old print. (Author)

deliberately and knowingly. They include landforms produced by constructional activity (tipping, grading, terracing, ploughing); by excavation (digging, cutting, mining, blasting, cratering, trampling and churning); and by hydrological interference (flooding, damming, dredging, bypassing, draining, etc.).

There are also those landforms which are produced indirectly and inadvertently. These are often less easy to recognize, not least because they tend to involve not the operation of a new process or processes, but the acceleration of natural processes. However, such changes are a crucial aspect of anthropogeomorphology, involving such processes as accelerated erosion and sedimentation as a result of land-use changes such as deforestation and urbanization; land subsidence as a result of mining, drainage and the like; and slope failures produced by the loading, undercutting, shaking or lubrication of slope materials.

In addition there are occasions where, through a lack of understanding of the operation of processes and the links between different processes and phenomena within complex systems, humans may deliberately and directly alter landforms and processes and thereby set in train a series of events which were not anticipated or desired.

Landforms produced by excavation

Sherlock (1931) made a quantitative attempt to provide a general picture of the importance of excavation in the creation of the British landscape. He estimated (see table 9.2) that up until 1913 humans had excavated around 31 billion m^3 of material in their pursuit of economic activities. On the basis of his calculations he was able to state (p. 333) that, 'at the present time, in a densely peopled country like England, Man is many more times more powerful, as an agent of denudation, than all the atmospheric denuding forces combined'. It must be remembered that, if that were true in 1913, it is even more true today; at the time when Sherlock wrote earth-moving equipment was only in a rudimentary state of development. Another notable change since Sherlock wrote has been in the widespread adoption of concrete as a construction material, with all the attendant demands that makes for the supply of aggregates. Demand for these materials in the UK grew from 20 million tonnes per annum in 1900 to 276 million tonnes in 1973, an increase in per capita consumption from 0.6 tonnes per year to about 5 tonnes per year (Jones, 1983).

Table 9.2 Total excavation of material in Great Britain until 1913

Activity	Approximate volume (m^3)
Mines	15,147,000,000
Quarries and pits	11,920,000,000
Railways	2,331,000,000
Manchester Ship Canal	41,154,000
Other Canals	153,800,000
Road Cuttings	480,000,000
Docks and harbours	77,000,000
Foundations of buildings plus street excavation	385,000,000
Total	30,539,954,000

Source: Sherlock (1931, p. 37)

In Britain, excavational activities have some antiquity. In the Breckland of East Anglia Neolithic peoples used antler picks to dig a cluster of deep pits in the Chalk at Grimes Graves to obtain non-frost-shattered flint for making stone tools. There were other notable flint mines on the South Downs behind Worthing, where there are extensive remains at Cissbury, Blackpatch, Findon and Harrow Hill.

In medieval times (between the ninth and fourteenth centuries AD) peat diggers excavated over 25 million m³ of peat to create the depressions in which the Norfolk Broads lie (Lambert et al., 1970; Moss, 1984), although previous workers (e.g. Jennings, 1952) had mistakenly believed them to be natural lakes. The need to excavate chalk marl for improving acidic, light sandy soils was practised widely in Britain, especially in the eighteenth century, and Prince (1962, 1964) has shown that many of the 27,000 pits and ponds in Norfolk resulted from this activity. Livingstone (1980) describes some of the pits, including 'dene holes' found in the chalklands of Kent and, while feeling able to attribute many of them to marling, stresses that others may result from other human (e.g. dew ponds) or natural processes (e.g. solution).

Moats and associated ditches are another ancient excavational feature of the landscape. More than 5,000 are known from England and 140 from Wales (most of which are along the English border). They are especially numerous in the southern half of East Anglia, and most were constructed in the twelfth or thirteenth centuries (Rackham, 1986, pp. 360–4).

Landforms produced by construction

The process of constructing mounds and embankments and the creation of dry land where none previously existed is longstanding. In Wiltshire, the great mound at Silbury Hill dates back to c.4,400 BP and is roughly contemporary with some of the Egyptian pyramids. However, probably the most important constructional landforms are those resulting from the dumping of waste materials, especially those derived from mining. It has, for example, been calculated (Richardson, 1976) that there are at least 2,000 million tonnes of shale waste lying in pit heaps in the coalfields of Britain.

Cities produce great amounts of wastes which need to be disposed of; many of the features created by excavation in one generation are therefore filled in by another. Phenomena like water-filled hollows, whether they be natural or created by marl-diggers, often are both wasteful of land and suitable locations for the receipt of waste. Watson (1976), for example, has mapped the distribution of hollows in the lowlands of south-west Lancashire and north-west Cheshire as they were represented on mid-nineteenth century topographic maps. He found that when this distribution is compared with the present-day one for the same area (figure 9.1) it is evident that a very substantial proportion of the holes has been filled in and obliterated, with only 2,114 out of 5,380 remaining. Hole densities had fallen from 121 km ^{2} to 47 km ^{2}.

Much accumulation takes place within cities themselves, so that the surface level of towns tends to rise as the town grows older: buildings decay or burn and are rebuilt, litter, house and trade waste accumulate, cesspits become filled, bodies are buried, etc. According to Sherlock (1931, p. 122), on average cities deposit a foot of debris per century. In the City of London he estimates that there is an average of 3–5 m of debris or made ground and a maximum of 8 m.

Other industries have also produced waste dumps; for example, there are many mounds and embankments associated with the coastal saltmaking industries, especially in low-lying areas like the Essex estuaries, the Humber marshes, the Wash, the Solway Firth and Morecambe Bay (Pattison and Williamson, 1986). The mounds of the north Kent coast, which are especially common on the marshes that border Sheppey and the Isle of Harty, and those around Graveney and at Seasalter Level (sic) are more than 50 m across and over 4.5 m in height (Holmes, 1981).

Accelerated water erosion of soils

It is difficult to assess the importance and extent of soil erosion in Britain for there is 'almost complete absence of measurements of the rate at which it occurs' (Morgan, 1977, p. 1). None the less there is evidence that in some parts of the country accelerated soil erosion is a problem and that various land-use changes may have exacerbated the problem. Such changes include the ploughing up of ancient pastures; the removal of field boundaries; the increased adoption of autumn sown cereals; the ploughing and drainage of

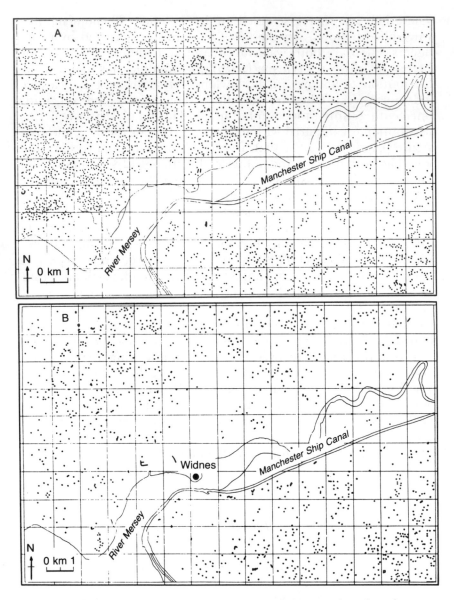

Figure 9.1 The distribution of pits and ponds in a portion of north-western England (modified from Watson, 1976): (A) in the mid-nineteenth century; (B) in the mid-twentieth century. The pits were largely a result of marl digging and in recent decades many have been filled in.

Table 9.3 Annual rates of soil loss (tonnes per hectare) under different land-use types in eastern England

Plot	Splash	Overland flow	Rill	Total
1 Bare soil				
Top slope	0.33	6.67	0.10	7.10
Mid-slope	0.82	16.48	0.39	17.69
Lower slope	0.62	14.34	0.06	15.02
2 Bare soil				
Top slope	0.60	1.11	—	1.71
Mid-slope	0.43	7.78	—	8.21
Lower slope	0.37	3.01	—	3.38
3 Grass				
Top slope	0.09	0.09	—	0.18
Mid-slope	0.09	0.57	—	0.68
Lower slope	0.12	0.05	—	0.17
4 Woodland				
Top slope	—	—	—	0.00
Mid-slope	—	0.012	—	0.012
Lower slope	—	0.008	—	0.008

Source: Morgan (1977)

upland areas for re-afforestation (Boardman and Robinson, 1985); and greater recreational pressures (Liddle, 1975).

Morgan's (1977) study of soil erosion on the sandy soils of mid-Bedfordshire indicates clearly both the importance of land use and the seriousness of the process. Rates of soil loss under bare soil on steep slopes can reach 17.69 t ha^{-1} y^{-1}, compared with 2.39 under grass and nothing under woodland (table 9.3). Morgan believes (p. 35) that:

the rate of soil loss on bare ground is sufficiently high to be considered unacceptable especially in view of the thinness of the top soil. There is some evidence to suggest that present rates fall short of the potential maximum rate and that if the resistance of the soil to detachment decreases erosion may increase still further. Overland flow is the most important erosive process, accounting for over 90 per cent of the soil loss from bare ground.

Morgan (1986) has prepared a map of England and Wales (figure 9.2) to show those areas which are potentially vulnerable to soil erosion. It is based on knowledge of soil characteristics, wind velocities and rainfall erosivity in particular areas. He recognized the following areas as having an erosion risk:

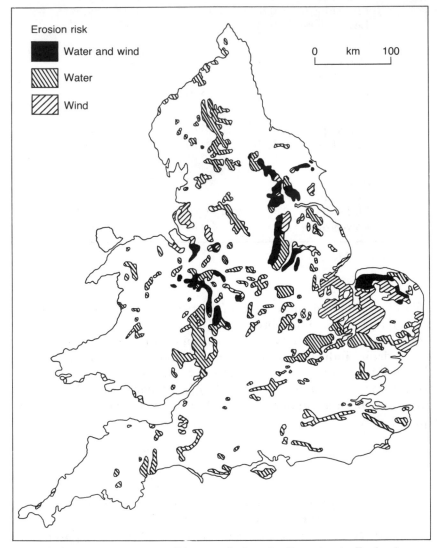

Figure 9.2 Areas susceptible to agricultural soil erosion in England and Wales (after Morgan, 1986, figure 4.17).

1 The Vale of York, parts of Nottinghamshire and North Norfolk: sandy and coarse loamy soils are prone, when used for arable farming, to both water and wind erosion.
2 Much of the Midlands, Welsh Marches (particularly Shropshire and Herefordshire) a belt extending from Berkshire north eastwards into Bedfordshire, Cambridgeshire and Suffolk, parts of south Devon, south Somerset, Dorset, Isle of Wight, east Hampshire and the North and South Downs: sandy, sandy loam, loamy and silty soils are prone, when used for arable farming, to water erosion.
3 North Norfolk, east Suffolk, parts of Lincolnshire, Yorkshire and Nottinghamshire and areas of lowland peat soils in the Fens, western Lancashire and Somerset: sandy and sandy loam soils are prone, when used for arable farming, to wind erosion.
4 Extensive areas of the Pennines and the Welsh Mountains, and smaller areas in the Lake District, Dartmoor and Exmoor: blanket peats are vulnerable to water, especially gully, erosion.
5 Areas of coastal sands: subject to wind erosion.

Most water erosion in England and Wales tends to occur in late winter and spring (Evans and Cook, 1986), with a notable peak in April and May (figure 9.3). The reasons for this timing include the fact that many fields still have only a limited crop development at that time. Also, the soils have been saturated for much of the winter and as the season progresses the surface becomes smoother as the soil slakes, so facilitating run-off. In addition, spraying the ground with herbicide prior to or just after the emergence of crop seedlings is a common practice and this leaves compacted wheel tracks in which rainfall collects, runs off and erodes.

Rates of erosion can be accelerated by the ditching of upland areas to improve drainage conditions for afforestation schemes. Artificial drainage networks thus created may have densities of 200 km^{-2}, 60 times the natural figure (Robinson and Blyth, 1982). Newson has undertaken studies of disturbed catchments in mid-Wales (Newson, 1981) and has shown that the yields of gravel bedload are modest in undisturbed grassland catchments but greatly elevated (table 9.4) in afforested and moorland catchments that have been ditched. The ditching disturbs the dense grass sod of upland grasslands, exposes underlying drift deposits to the effects of stream run-off, and may enhance peak flows (Robinson, 1986). Such ditching also causes an increase in suspended sediment yields, and in a study of the small Coalburn catchment (40 km NE

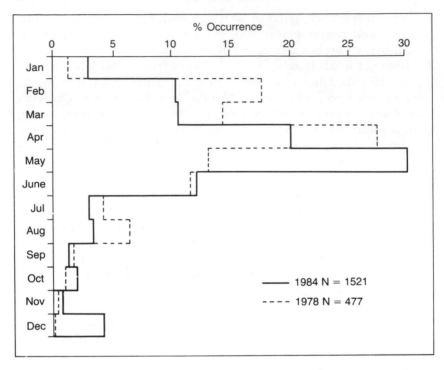

Figure 9.3 Months when water erosion of sample fields took place in England and Wales in 1978 and 1984. Note the high level of occurrences in spring and early summer (after Evans and Cook, 1986, figure 4).

of Carlisle) Robinson and Blyth (1982) found that forestry drainage operations can locally increase average stream sediment loads at least 50-fold for short periods, and they calculated that 120 tonnes km^{-2} of sediment was lost in a five-year period as a result of the drainage. The long-term effect was probably to quadruple annual yields from 3 to 12 tonnes $km^{-2} y^{-1}$.

These studies are of fundamental interest because traditionally the presence of a forest has been regarded as a dampening influence on sediment loss, with accelerated erosion generally following deforestation. Such tradition is largely based on studies of established forests in countries where forestry drainage is not common. In Britain, for climatic and pedological reasons, drainage is practised and it is ditch erosion that can create the high erosion rates as a result of afforestation.

Table 9.4 Gravel bedload sediment yield
from small catchments in mid-Wales

Catchment	Annual yield $(m^3 \ km^{-2})$
Grassland undisturbed	
Cyff	2.5
Cownwy	2.5
Peny y Banc	9.9
Maesnant	1.1
Iago	1.2
Forested, ditched	
Tanllwyth	8.4–307.7
Groes	44.4
Marchnant	30.9
Moorland, ditched	
Bugeilyn	57.1
Hengwm	2.0–17.9
Llyn Pen Rhaidr	3.5–38.8

Source: Newson (1971, table 3.1, p. 78)

Another cause of accelerated erosion is building activity. Disturbance during the construction phase can add substantial amounts of sediments to streams, although the data on this effect in Britain are sparse. Walling (1979) studied the effects of partial suburbanization of a rural catchment near Exeter in Devon and, although building activity only influenced 25 per cent of its area, sediment loads increased five to tenfold. However, the effects of construction work are not perpetual and, as the disturbance ceases, as roads are surfaced and as gardens and lawns are established the rates of erosion fall dramatically.

Thus although the low rainfall erosivity regime experienced in Britain has often encouraged the view that water erosion of arable and other soils is slight, there is increasing evidence that on susceptible soils potentially erosive storms are quite frequent. Fuller and Reed (1986) found that in Shropshire very serious erosion could be achieved by falls with an intensity of > 10 mm h^{-1}, especially if they occur in late spring or early summer when the crop cover has yet to develop fully.

One consequence of water erosion is the formation of sunken lanes or holloways. Centuries of traffic on unpaved roads loosens the surface and prevents vegetation from holding it, and rain washes away the debris. This phenomenon is common on the Upper Greensand around Midhurst in the Sussex Weald and on the Lower Greensand in Wiltshire and Dorset (Rackham, 1986, pp. 278–9).

Peat erosion

In has been calculated that over 10,000 km^2 of the land surface of Great Britain is peat-covered. Deposits of blanket peat, through their influence on patterns of streamflow and erosion in upland catchments, are of major geomorphological significance. However, some areas of the blanket peats are severely eroded, notably in the Pennines (Bower, 1961; Conway, 1954); Tallis (1985a) has estimated that around three-quarters of the blanket peat in the Peak National Park is affected. The erosion problem is especially serious at higher altitudes (especially above about 550 m OD), producing pool and hummock topography, areas of bare peat, and incised gullies. Terms of Norse origin (hagg, grough, grain) are used for certain of these erosion features.

Bare peat surfaces are unstable and abundant evidence of this instability can be seen in the field (Tallis, 1985b, p. 318):

> Finely-divided peat, washed downslope, accumulates against objects in its path, such as walls and wire fences; even clumps of bilberry may be partly buried by this peat wash; and residual patches of snow, persisting after snow-melting in spring, quickly acquire a superficial layer of peat particles. In areas of scour, established individual plants come to be raised up on peat pedestals several centimetres high, as the surrounding peat surface is gradually lowered. After heavy rain and during snow-melt, turbid water rushes down the drainage gullies, carrying away peat in suspension and solution. In dry weather the superficial layers of peat shrink and crack, and may blow away as dust, while the cracks extend downwards.

Rates of erosion on such surfaces in the southern Pennines range from 5 to 34 mm y^{-1} (Tallis, 1985b, p. 319).

Some of the observed peat erosion may be an essentially natural process, for the high water content and low cohesion of undrained peat masses make them inherently unstable. Moreover, the instability must normally become more pronounced as peat continues to accumulate, leading to bog slides and bursts round margins of expanded peat blankets. Conway (1954) suggested that an inevitable end-point of peat build-up on high altitude, flat or convex surfaces is that a considerable depth of unconsolidated and poorly humidified peat overlies denser and well-humified peat, so adding to the instability. Once a bog burst or slide occurs, these lead to the formation of drainage gullies which extend back into the peat mass, slumping-off the marginal peat downslope, and leading to the drawing off of water from the pools of the hummock and hollow topography of the watershed.

Tallis (1985a) believes that there have been two main phases of erosion in the area. The first, initiated 1,000–1,200 years ago, may have been caused by natural instability of the type outlined above. However, there has been a second stage of erosion, initiated 200–300 years ago, in which miscellaneous human activities appear to have been important. Radley (1962) suggested that among the pressures that had caused erosion were heavy sheep grazing, regular burning, peat cutting, the digging of boundary ditches, the incision of pack horse tracks, and military manoeuvres during the First World War. Other causes may include footpath erosion and severe air pollution (Tallis, 1965).

Peat erosion also occurs in other parts of upland Britain. Thomas (1956) gives examples from the Brecon Beacons of Wales and his description of erosion mirrors that of Tallis from the Pennines. He describes gullies and eroded mounds or haggs, and refers to the spread of the erosive process (p. 103):

When dry easterly winds of anticyclonic spells predominate, aerial transportation of particles of crumbly peat exposed in the gully banks is often evident. Prolonged rainfall or sharp downpours are liable to produce local changes of configuration with the evolution of branch gullies whilst frost-heaving undoubtedly assists in the formation of fresh peat exposures. With advancing maturity of the gullying cycle the original peat blanket is more and more reduced so that in the final stages a few rectangular piles of peat with some surviving tussocks of cotton grass (*Eriophorum vaginatum*) are surrounded by a thin layer of redistributed peat partially covering the solid or subsoil base.

There is some evidence that the blanket peats of South Wales have degenerated as a result of contamination by particulate pollution during the industrial revolution (Chambers et al., 1979).

Wind erosion

There are some soil types in Britain which may be inherently susceptible to wind erosion. These include drained organic soils in areas like the Fenland, and relict Quaternary coversands (see p. 107) in areas like the Breckland, East Yorkshire and Lincoln-shire (figure 9.4). Ever since the 1920s dust storms ('soil blows') have been recorded in these areas (Arber, 1946; Radley and Sims, 1967; Robinson, 1968).

The increasing incidence of soil blows results from changing agricultural practices, including the substitution of artificial fertilizers for farmyard manure; a reduction in the process of 'claying' whereby clay was added to the peat to stabilize it; the removal of hedgerows to facilitate the use of bigger farm machinery; and, perhaps most importantly, the increased cultivation of sugar beet. This crop requires a fine tilth and, compared with other crops (Pollard and Miller, 1968), tends to leave the soil relatively bare in early summer (a time when there can be long spells of dry windy weather in eastern England).

Not all wind erosion is necessarily very recent. Windbreaks were planted in the Breckland in the last century to stop sandblowing, a problem which was greatly accelerated by severe rabbit infestation. In the seventeenth century blow-outs and dune migration had occurred blocking tracks and even overwhelming the village of Santon Downham (Clarke, 1925). Similarly, Daniel Defoe (1724–6) penned his experiences of 'the horrid and frightful' Bagshot Heath (1971 edition, p. 156):

Figure 9.4 The distribution of wind erosion events, 'soil blows', in Lincolnshire, March 1968 (after Robinson, 1968).

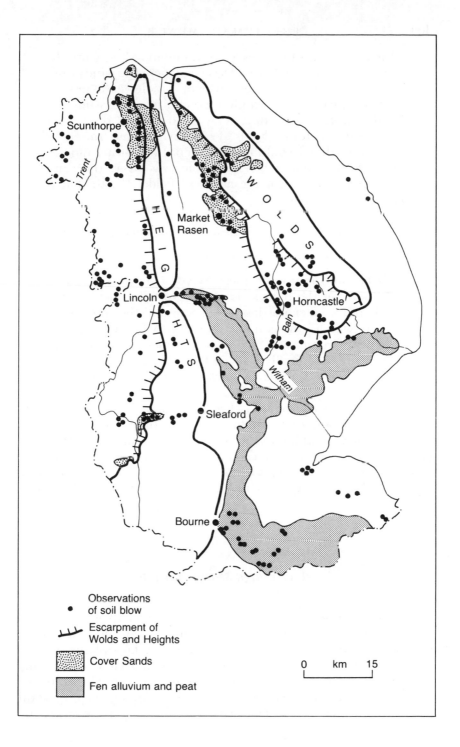

Scunthorpe

Trent

WOLDS

HEIGHTS

Market
Rasen

Lincoln

Horncastle

Bain

Witham

Sleaford

Bourne

Observations
of soil blow

Escarpment of
Wolds and Heights

Cover Sands

Fen alluvium and peat

0   km   15

> Much of it is a sandy desert, and one may frequently be put in mind here of Arabia Deserta, where the winds raise the sands, so as to overwhelm whole caravans of travellers, cattle and people together, for in passing this heath, in a windy day, I was so far in danger of smothering with the clouds of sand, which were raised by the storm, that I could neither keep it out of my mouth, nose or eyes; and when the wind was over, the sand appeared spread over the adjacent fields . . . so that it ruins the very soil.

In 1884 Clement Reid (p. 167) described the effects of soil blowing in Norfolk:

> It is not uncommon on these Wolds (Lincolnshire) and still more on the light soils of the Norfolk coast, for the whole of the finer portion of the soil, and the seed, to be blown away by the equinoctial gales. A few years ago a field near Cromer was thrown three times in succession in one spring, and finally was left fallow, as the whole of the soil was banked up like a snow-drift against the hedge.

Wind erosion of soils shows a clear seasonality. Erosion generally occurs in spring and early summer (March–June), in fields sown to sugar beet or vegetables and which have a dry smooth surface. As table 9.5 shows, 'blows' which caused moderate or severe damage to crops in two areas of Cambridgeshire and Nottinghamshire occurred about one year in two for each month from March to May.

Accelerated sedimentation

The erosion consequent upon major land-use change has created accelerated sedimentation in favourable topographic situations, most particularly in lake basins and on valley floodplains.

Various studies have been undertaken with a view to assessing the importance of changes in sedimentation rate caused by humans at different times in the Holocene (table 9.6). Among the important events that have been identified are initial land clearance by

Table 9.5 Frequency of damaging winds (years in ten, 1968–77)

	Mepal Cambridgeshire		Gleadthorpe Nottinghamshire	
	Severe blow	Moderate blow	Severe blow	Moderate blow
March	1	2	1	4
April	1	4	1	4
May	2	4	4	1
June	0	2	1	0

Source: undated Ministry of Agriculture, Fisheries and Food data in Evans and Cook (1986, p. 30, table 1)

Mesolithic and Neolithic peoples; agricultural intensification and sedentarization in the Late Bronze Age; the widespread adoption of the iron plough in the early Iron Age; and the introduction of Viking settlements and sheep farming.

A core from Llangorse Lake in the Brecon Beacons (Jones et al., 1985) provides excellent long-term data on changing sedimentation rates:

Period (y BP)	Sedimentation rate (cm 100 y^{-1})
9000–7500	3.5
7500–5000	1.0
5000–2800	13.2
2800–AD 1840	14.1
c.AD 1840–present	59.0

The 13-fold increase in rates after 5,000 BP seems to have occurred rapidly and is attributed to initial forest clearance. The second dramatic increase of more than fourfold took place in the last 150 years and is a result of agricultural intensification.

In the last two centuries, rates of sedimentation in lake basins have changed in different ways in different basins according to the differing nature of economic activities in catchments.

Some data from various sources are listed for comparison in table 9.7. In the case of the Loe Pool in Cornwall rates of sedimentation were high while mining industry was active, but fell dramatically when mining was curtailed. In the case of Scaswood

Pool in Warwickshire, a dominantly agricultural catchment area, the highest rates have occurred since 1978 in response to various land management changes (e.g. larger fields, continuous cropping and increased dairy herd size).

In other catchments, pre-afforestation ploughing may have caused sufficient disturbance to cause accelerated sedimentation. For example, Battarbee et al. (1985) looked at sediment cores in the Galloway area of south-west Scotland and found that in Loch Grennoch the introduction of ploughing in the catchment caused an increase in sedimentation from 0.2 cm y^{-1} to 2.2 cm y^{-1}.

Table 9.6 Accelerated sedimentation in prehistorical and historical times

Location	Source	Evidence and Date
Howgill Fells	Harvey et al. (1981)	Debris cone production following tenth century AD introduction of sheep farming
Upper Thames basin	Robinson and Lambrick (1984)	River alluviation in Late Bronze Age and early Iron Age (post-3,000 BP)
Lake District	Pennington (1981)	Accelerated lake sedimentation at 5,000 BP as result of Neolithic agriculture
Mid-Wales	Macklin and Lewin (1986)	Floodplain sedimentation (Capel Bangor unit) on Rheidol as a result of early Iron Age sedentary agriculture
Brecon Beacons	Jones et al. (1985)	Lake sedimentation increase after 5,000 BP at Llangorse due to forest clearance
Weald	Burrin (1985), Scaife and Burrin (1987)	Valley alluviation from Neolithic onwards until early Iron Age
Bowland Fells	Harvey and Renwick (1987)	Valley terraces at 5,000–2,000 BP (Bronze or Iron Age settlement) and after 1,000 BP (Viking settlement)
Southern England	Bell (1982)	Fills in dry valleys. Bronze Age and Iron Age

Table 9.7 Data on rates of erosion and sedimentation in lakes in the last two centuries.

Dates	Activity	Rates of erosion in catchment as determined from lake sedimentation rates (tons $km^{-2} y^{-1}$)
Loe Pool (Cornwall)[a]		
1860–1920	Mining and agriculture	174
1930–36	Intensive mining and agriculture	421
1937–8	Intensive mining and agriculture	361
1938–81	Agriculture	12
Seeswood Pool (Warwickshire)[b]		
1765–1853		7.0
1854–80		12.2
1881–1902		8.1
1903–19		9.6
1920–25		21.6
1926–33		16.1
1934–47		12.7
1948–64		12.0
1965–72		13.9
1973–7		18.3
1978–82		36.2

[a] From O'Sullivan et al. (1982)
[b] From Foster et al. (1986)

Ground subsidence

Natural processes (e.g. limestone solution) can cause ground subsidence in the absence of humans, but ground subsidence can be created or accelerated anthropogenically in a variety of ways: by the transfer of subterranean fluids; by the disruption of permafrost to produce thermokarst; by the removal of solids through underground mining or by dissolving solids and removing them in solution; and by the compaction or reduction of sediments because of drainage and irrigation. In Britain it is the last two of these four types that is of greatest significance.

Mining subsidence led to court cases in England as early as the fifteenth century, although its importance varies according to such factors as the thickness of seam removed, its depth, the width of working, the degree of filling with solid waste after extraction, the geological structure, and the method of working adopted (Wallwork, 1974). In general terms, however, the vertical displacement by subsidence is less than the thickness of the seam being worked, and decreases as the depth of mining increases. This is because other overlying strata collapse, fragment and fracture, so that the mass of rock fills a greater space than it did when naturally compacted. Consequently, the surface expression of deep-seated subsidence may be equal to little more than one-third of the thickness of the material removed.

Coal-mining subsidence may produce depressions which disrupt surface drainage and cause flooding to give permanent lakes as in the Wigan–Leigh area of Lancashire; in the valleys of the Aire and the Calder in the vicinity of Castleford and Knottingley; in the Anker valley of east Warwickshire; and in South Staffordshire. The degree of local subsidence may be appreciable, with reports of as much as 12 m in parts of Staffordshire (Sherlock, 1931, p. 76).

Subsidence in coal-mining areas can be accompanied by small-scale earth tremors. In North Staffordshire, where they are known locally as 'Goths', or 'Bumps', the Trent Vale area has recently been subject of a plethora of small events, some of which caused damage to property.

Extreme examples of subsidence have also been associated with the exploitation of Triassic salt deposits, especially in Cheshire. The seams are often thick (up to 30 m) and at shallow depths (c.70 m below the surface), so the subsidence risk is high. Furthermore, in contrast to coal, rock salt is highly soluble in water, so that the flooding of mines may cause additional collapse. The mining of salt and pumping of brine has led to the formation of various lakes, locally called 'flashes' (Wallwork, 1956), and caused the railway line between Crewe and Manchester to subside 4.9 m between 1892 and 1956 at Elton (Wallwork, 1960).

Locally important examples of ground subsidence have been associated with the draining of organic soils. The lowering of the water table makes peat susceptible to oxidation and deflation, so that its volume decreases. The measurements at Holme Fen in the East Anglian Fens indicate that 3.8 m of subsidence occurred between 1848 and 1957 (Fillenham, 1963) with the fastest rate occurring soon after drainage had been initiated. The present rate

Plate 32. The drainage of the East Anglian Fens has caused the land to subside and the peat to waste away. This reveals old silty creek systems (roddens) which emerge as lighter coloured ridges rising above the peat. (Cambridge University Collection: copyright reserved)

averages about 1.1–1.4 cm y^{-1} (Richardson and Smith, 1977; Hutchinson, 1980b). At its maximum the natural extent of peat in the area was 1,750 km². Now only about one-quarter (430 km²) remains. Only small areas of original surface fen are extant (e.g. Wicken Sedge Fen and St Edmunds Fen) and these, although themselves probably lowered 3 m over the centuries, still stand 2 m above the surrounding drained land.

Land drainage has also caused shrinkage on the reclaimed coastal marshlands of Romney Marsh. Dendritic creek systems, the beds of which were composed of sandy materials, were cut into areas of peat. However, following drainage the peat shrank much more than the sandy creek material and, as a consequence, relief inversion occurred causing the creek channel sands to form dendritic ridges (figure 9.5).

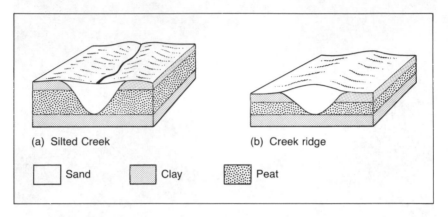

(a) Silted Creek   (b) Creek ridge

☐ Sand   ☐ Clay   ▨ Peat

Figure 9.5 Creek ridge development associated with peat wastage (from Curtis et al. 1976, figure 13.13a).

There is very little evidence in Britain that groundwater abstraction has been a very important cause of ground subsidence. Even in London, where pumping lowered groundwater levels in the underlying Chalk aquifer by more than 60 m, ground subsidence between 1865 and 1931 amounted to no more than 0.06–0.08 m.

Accelerated weathering

It is likely, although not generally proven, that human activities have accelerated rates of rock weathering. There is considerable evidence for severe weathering of building stones in polluted urban environments and some monitoring is in progress (see, for example, Jaynes and Cooke, 1987) but the evidence for accelerated decay of natural stones in situ in rural locations is very much more sparse.

It can be believed intuitively that rates of weathering should have changed, for since the turn of the century the UK has been emitting large quantities of SO_2 and NO_2 (see table 9.8). These materials undergo oxidation to sulphate aerosol and gaseous nitric

Table 9.8 UK emissions of SO_2 and NO_2 since 1900 (in millions of tonnes per year) from all sources

	SO_2	NO_2
1900	2.8	0.68
1910	3.2	0.69
1920	3.2	0.72
1930	3.2	0.72
1940	3.6	0.82
1950	4.6	0.99
1960	5.6	1.35
1970	6.0	1.64
1980	4.67	1.79
	(2.87 from power stations)	(0.85 from power stations)
		(0.49 from vehicles)
Increase (%)	167	263

Source: Watt Committee on Energy Report No. 14 (1984) tables 1.1 and 1.2

acid respectively and are deposited to give acidic precipitation. Much of these materials comes from power stations and from vehicles.

Natural precipitation has a pH value of around 5.67. It is slightly acid because of the presence of dissolved carbon dioxide which reacts with water to produce carbonic acid. Much of Britain now has average pH values for precipitation of less than 4.5, and rainfall with a pH of 3.0–4.0 appears to be relatively common. The pH of cloud water in the Pennines can be as low as 2.2–2.5, while in the great London smog of 1952 the pH was between 1.4 and 1.9.

Channel changes

Engineering works involving diversion, straightening, conduiting and bank protection cause channel modifications as does the canalization of rivers for navigation purposes. The embankment of estuaries like the Thames, and the formation of major drainage lines in the Fens, are examples of deliberate and purposeful change.

The history and extent of channelization has been described by Brookes (1988), and figures 9.6 and 9.7 show the pattern of channel modification that has taken place in different areas between 1930

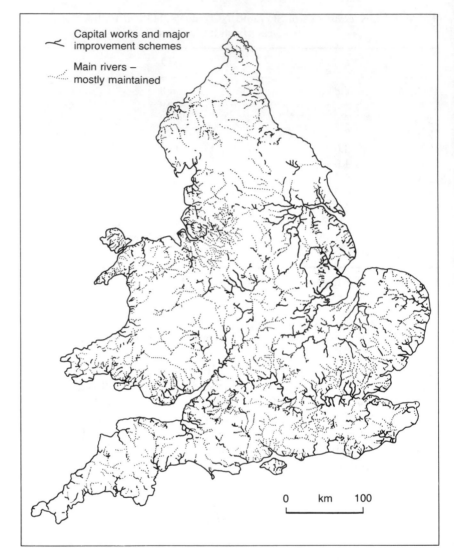

Figure 9.6 Rivers channelized in England and Wales from 1930 to 1980 (after Brookes et al., 1983).

and 1980. The percentage of main river which is channelized varies from 41 per cent in the London area to 12.3 per cent in the North West and Thames Water Authority areas.

Rather less obvious but no less important are the various indirect

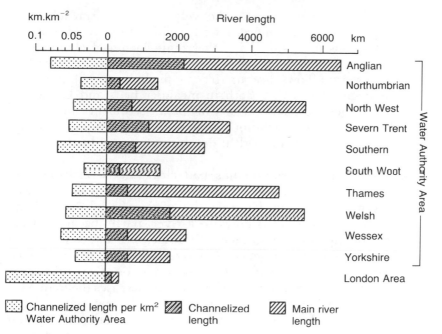

km.km⁻²   River length

Channelized length per km² Water Authority Area / Channelized length / Main river length

Figure 9.7 Lengths of channelized river in Water Authority areas in England and Wales. The works included those undertaken by Catchment Boards, River Boards, River Authorities and Regional Water Authorities between 1930 and 1980 (after Brookes et al., 1983).

effects on channel behaviour of human interference with the hydrological cycle. Among the important influences are reservoir construction, interbasin water transfers, urbanization and the disposal of mining sediment.

The overall effect of the creation of a reservoir by the construction of a dam is to lead to a reduction in downstream channel capacity. As Petts (1979) has shown (see table 9.9) this seems to amount to between about 30 and 70 per cent. However, the magnitude and nature of the aggradation response to flow regulation varies between as well as along rivers (Ferguson, 1981, p. 123):

> Many coarse-bed reaches with well vegetated banks do not appear to have changed since reservoir construction, presumably because reduced peak flows have not reached the competent threshold and are passively accommodated

within a misfit channel. Actively meandering reaches appear to adjust more readily with fairly consistent reductions in width. The other main locum of channel change is below tributary junctions where reduced mainstream flows are unable to flush away tributary sediment inputs . . . Both in meander belts and below tributaries aggradation is mainly of fine sediment trapped by encroaching vegetation; some bedload deposition may occur but in general width: depth ratios are reduced.

The spread of urbanization has had clear effects on catchment hydrology and stream behaviour, and these in turn have caused changes in channel form. Urbanization of river basins results in an increase in the peak flood flows in a river, partly because of the presence of impermeable surfaces of tarmac, tiles and concrete and partly because of the presence of sewers and storm drains. Given that stream channel morphology depends on their discharge characteristics, and given that urbanization increases the frequency of discharges which fill the channel, erodible materials will be eroded so as to enlarge the channel. Gregory and Park (1976) found that channel capacities in the town of Catterick, North Yorkshire, were up to 150 per cent higher than in rural tributaries of the same streams, while Hollis and Luckett (1976) found a 70 per cent increase in channel capacity in West Sussex. In Manchester, suburbanization of the River Bollin has led to a reduction in channel sinuosity as a response to the suburban regime (Mosley, 1975).

Afforestation may also affect channel form. Work by Murgatroyd and Ternan (1983) in the Narrator Brook catchment of south-west Dartmoor showed marked differences between forested and non-

Table 9.9 Channel capacity reduction below British reservoirs

River	Dam	Channel capacity loss (%)
Tone	Chatworthy	54
Meavy	Burrator	73
Nidd	Angram	60
Burn	Bwn	34
Derwent	Ladybower	40

Source: Petts (1979, table 1)

forested channel reaches: the forested reaches were wider, shallower and less sinuous. These changes in channel form result from active bank erosion within the forest with coarse material being deposited within the channel as mid-channel bars and point-bars. Such active bank erosion can be attributed primarily to the fact that under the forest the thick grass turf with its dense network of fine roots is suppressed. Secondly, within the forest, the river has to bypass log jams and debris dams in the stream channel.

The disposal of mining waste may have implications for river channel behaviour. The introduction of large amounts of coarse mining waste from the lead and zinc mines of mid-Wales in the nineteenth century was to cause channel braiding and aggradation (by as much as 3 m) in rivers like the Ystwyth (Lewin et al., 1983). In Cornwall and Devon, the exploitation of deposits of china clay has caused large amounts of clay waste to be introduced into channels, causing silting of estuaries and exacerbated flood risk (Cominetti, 1986). Surveyed dimensions of polluted and unpolluted channels (Richards, 1979) shows a significant difference in downstream trend, with the polluted channels tending to be narrower and deeper. The extra clay gives greater bank cohesion so that a steep face can be maintained to constrain the channel.

One particular type of channel modification, more extensive in the past than it is now but still widely visible in the landscape today, is the water meadow. Two main types occur in Britain and are commonly distinguished by the terms 'catch-work meadow' and 'flowing (or floated meadow'). The former is made by turning a spring or small stream along the side of a hill and thereby irrigating the land between the new cut and the original watercourse. This type is used frequently in the south-western peninsula and to some extent in Herefordshire. Floated meadows, formerly prevalent in the valley bottoms of Wessex (Moon and Green, 1940) and also in Bedfordshire and Northamptonshire, rely on the level of the river being raised by means of a weir provided with a large spill-way, and a battery of hatches, so that excess water may be run off at times of flood. From just above the weir a new channel (often called 'the main carrier') runs across the meadow at a higher level than the original watercourse and a series of smaller channels carry, and drain water back to the main river. The purpose of these channel modifications was to improve pasture growth, especially in the spring months, but the labour requirements, especially of the floated meadows, has led to their demise.

Deliberate channelization of part of a stream may lead to

unanticipated channel changes downstream. In England and Wales, channelization generally causes the enlargement of downstream cross-sections (Brookes, 1987), and this may be explained in terms of increased flood flow downstream from channelization works causing higher stream velocities. These, in turn, cause erosion, thereby increasing channel width and/or depth. The effect is more marked in upland than in low-energy lowland catchments, and it is width which increases preferentially to depth, probably because bed armouring or the presence of underlying rock restrict bed degradation.

Overall, however, the effects of humans on British rivers are less pronounced than those in many other countries. Their armoured gravel beds give them some stability and they do not carry massive sediment loads to cause large-scale sudden aggradation. Furthermore, they are small by world standards and drain a generally well-vegetated and relatively subdued landscape, so that for the most part they are naturally stable and well buffered against human interference (Ferguson, 1981, p. 125).

Accelerated slope instability

Many slopes in Britain (see p. 197) are subject to the various effects of mass movements: slope instability is widespread and most instability is caused by a particular combination of natural circumstances operating on susceptible materials. However, by means of either increasing the power of disturbing forces or decreasing the resistance of materials humans can accelerate the rate of slope failures.

Some mass movements are created by humans piling up waste, soil and rock into unstable accumulations that fail spontaneously. In 1966, at Aberfan in South Wales, a major and disastrous failure took place when a 180 m high coal-waste tip began to move as a debris flow. The tip had not only been constructed with a steep slope configuration, but it had been located on a spring line.

Other mass movements may have been accelerated by deforestation and agricultural activities. For example, Innes (1983) has demonstrated, on the basis of lichenometric dating of debris-flow deposits in the Scottish Highlands, that most of the flows have developed in the last 250 years, and he suggests that intensive burning and grazing may be responsible. Statham (1976) suggests sheep grazing as the cause of debris flows on the Black Mountains in South Wales.

Some of the most severe examples of accelerated slope instability occur on coasts. A classic example is provided by Folkestone Warren in Kent (figure 9.8), a large landslide complex developed where Chalk overlies Gault Clay. It lies to the east of Folkestone and the direction of littoral drift is from the west. Thus, the construction of large harbour works at Folkestone in the nineteenth century blocked the eastward movement of coastal shingle and the Warren became depleted in beach material. This made it more susceptible to undercutting and oversteepening by wave attack. A series of failures occurred in the nineteenth century and in 1915 a major failure disrupted the Folkestone–Dover railway line (Hutchinson et al., 1980). Remedial action, such as weighting the toe of the slip, has been employed to make the area less unstable.

Since Neolithic times, the adoption of ploughing, with ever more effective ploughs, has served to modify the rate of soil movement on slopes and has created suites of micro landforms. In pre-Roman times ploughing on steep slopes created small terrace-like features called lynchets, which occasionally have a tread height of as much as 8 m. In medieval times suites of parallel strip lynchets were created as population pressures forced cultivation of high angle slopes. Medieval strip lynchets tend to be longer than their prehistoric forebears, sometimes as much as 200 m, and where whole flights lie together on a hillside they are often linked by ramps which give access to the treads (Taylor, 1975).

The development of lynchets may still have been proceeding in the nineteenth century for their contemporary development was described by Scrope (1866) who reported that one of his tenants, by ploughing on a steep Wiltshire slope, had caused a two to three foot high step to be formed in just ten years (p. 294):

> Any one who lives in a neighbourhood where these banks occur may see them if not in course of formation from their beginning, yet growing yearly before his eyes; that is to say, wherever the slope *above* the bank is under arable cultivation. In this case as the course of the plough almost always follows the more or less horizontal trend of the surface, which is always the direction of the banks, the ridge of soil raised by the mould-board of the plough has everywhere a tendency, through the action of gravity upon it, to fall down-hill, never upwards. This down-hill

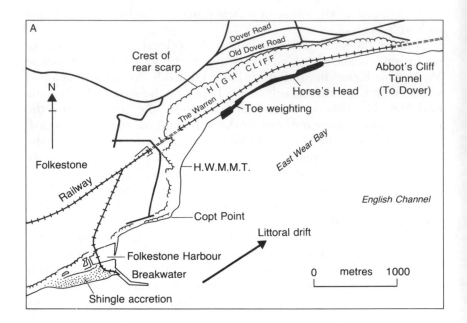

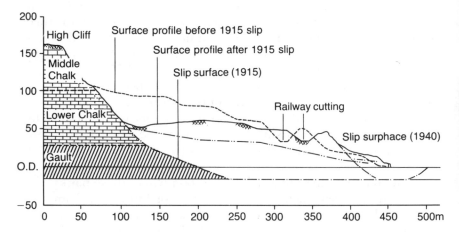

Figure 9.8 The Folkestone Warren landslip, Kent: (A) general location map showing the Warren, the breakwater of Folkestone Harbour, the direction of littoral drift, and the toe weighting structures; (B) cross-section showing the effects of the 1915 and 1940 slips.

tendency of the soil is greatly assisted by the wash of heavy rains.

Another major impact of ploughing is the development of ridge-and-furrow. Large areas, especially of the Midland lowlands, are striped by long, narrow ridges of soil, lying more or less parallel to each other and usually arranged in blocks of approximately rectangular shape. They were formed by ploughing with heavy ploughs pulled by teams of oxen; the precise mechanism has been described with clarity by Coones and Patten (1986, p. 154):

> The 'heavy' plough cut through the soil and turned the furrow-slice to the right. Consequently, if the field were simply ploughed up and down by working progressively across it from one side to the other, each double run would merely produce a tiny ridge overlying an unploughed ribbon of ground, the soil being thrown inwards from the two furrows on either side. This would have been a pointless exercise, so instead the field was ploughed as a number of separate units, each with its own central axis. Ploughing started along the centre line and worked gradually away from it, turning the soil inwards from both edges as the plough went along one side and back down the other. Every sod on each side was therefore turned the same way and laid against the previous one. It has been suggested that once the plough team had progressed to a point a certain distance from the original centre line, the distance it was obliged to walk along the headland to the other side for the return run was such that it was easier to start a new circuit, based upon the axis of an adjacent unit. Repeated ploughing of exactly the same units year after year gradually led to the formation of adjacent ridges, separated from each other by furrows produced where the furrow-slices were turned away from each other.

Coastal modification

The coast is an active environment both for geomorphological processes and for human activities. It is, therefore, not surprising that the one has frequently been affected by the other. As with other examples of the human impact it is relatively easy to identify deliberate modifications such as land reclamation by dumping, embanking and planting; coast protection by groynes and sea walls; dock and navigation channel excavation by dredging, etc. Sherlock (1931) gives many examples of such deliberate modifications of which the Tees estuary provides a wealth of different types (pp. 163 and 171):

> The most striking case of reclamation by direct filling up of the foreshore is probably to be found in the Tees estuary. Slag from the Middlesborough iron-furnaces was taken in vessels and dumped on the boundaries of the area to be reclaimed. When the surface of the slag-tips had risen above high-water mark, railways were laid down along them and the slag was then shot into the enclosed area until it had filled up. Some 4270 acres (1729 ha) have been filled or are in process of being filled up.
>
> * * *
>
> Up to 1885, the total amount dredged from the River Tees was 15,130,000 cubic yards [11,567,369 m^3] while more than 124,000 cubic yards [94,800 m^3] was removed by blasting. The river has also been straightened between Stockton and the sea by cutting through the meanders. One of these cuttings, that of Portrock, is 1100 yards long by 250 yards wide and 16 feet deep. For 7 miles above and 13 miles below Middlesborough, the river is kept in place by training-walls.

Deliberate coastal reclamation has a long history in Britain. The Romans reclaimed part of the Fens, and the Car Dyke is part of this work. Traces of sea wall have been found along the Essex coast and in Glamorgan (e.g. at Llantwit Major); while in Kent the Rhee Wall on Romney Marsh is also probably Roman. Further reclamation was carried out by the Saxons and Scandinavians before the Norman Conquest, and the process continued apace thereafter. By the seventeenth century a bank had been built along the south side of the Thames from Southwark to Greenwich; Canvey Island was reclaimed at this time (Wood, 1972).

Figure 9.9 shows the progress of reclamation since Saxon times around the Wash. Around 32,000 ha of salt marsh has been

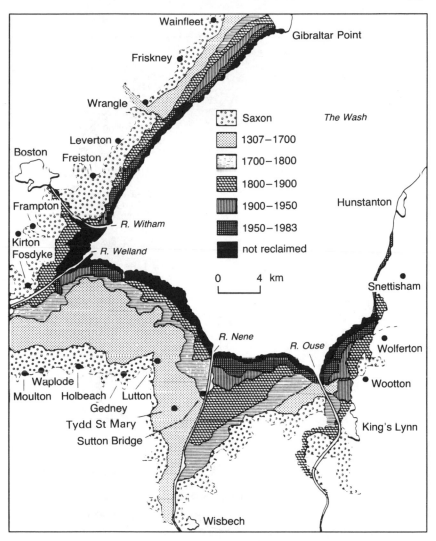

Figure 9.9 The progress of land reclamation around the Wash (after Doody and Barnett, 1987, frontispiece).

reclaimed for agriculture up to the present day (Doody and Barnett, 1987).

The vulnerable and generally lowlying coast of East Anglia, developed in erodible materials such as Pleistocene drift and Chalk, is now very heavily modified by a whole suite of defences, including clay banks, sea walls, revetments and groyne systems (figure 9.10).

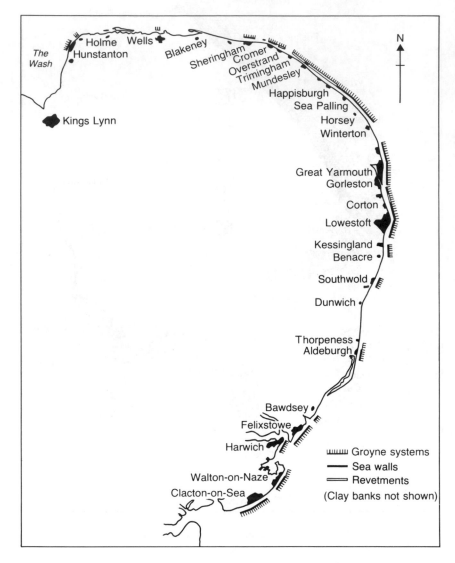

Figure 9.10 The coastline of East Anglia has been heavily 'protected' against coast erosion by long stretches of defence works, including groynes, sea walls and revetments. The cost, the aesthetics and the effectiveness of such structures has been questioned (after Steers, 1981, figure 7).

Nearly three quarters of the Norfolk cliffs are defended (Clayton, 1989). In the past, coastal management has tended to mean shore protection. Victorian engineers like John Coode and John Rennie sought to 'reclaim' land from the sea, although the prefix 're-' is often sadly misused in this context, as few of the areas in question had formerly been dry land (Carter, 1987).

However, the accidental changes wrought by human activities are also of great significance; in particular the modification of coastal sediment movement by the construction of engineering works or the removal of sediment. For example, the 'protection' of Norfolk cliffs, referred to above, has reduced the sediment supply to beaches downdrift to about 70–75 per cent of its natural level.

The physical removal of beach sediment, by quarrying or offshore dredging is potentially a very serious problem because in many parts of England and Wales beaches have received little sediment replenishment since sea-levels stabilized around 6,000 years BP (see p. 285). The classic example of beach mining causing cliff retreat is at Hallsands in Devon, where the mining of 660,000 tonnes of shingle in 1887 to provide material for the construction of the Plymouth dockyards caused the shore level to fall by around 4 m. The loss of this protective shingle soon resulted in cliff erosion to the extent of 6 m between 1907 and 1957. The village of Hallsands then became prone to wave attack and is now in ruins.

Engineering works, including groynes, piers and breakwaters, can interrupt longshore sediment movement, causing accumulation at one point and erosion at another. The example of the effect of the construction of harbour works at Folkestone on the cliffs at Folkestone Warren is discussed in the previous section. Other examples of severe erosion of this type from England are at Seaford in Sussex (caused by Newhaven breakwater), Lowestoft (caused by the pier at Gorleston), and at West Bay, Dorset (by the breakwater at the mouth of the harbour). In Wales there has been an anthropogenically caused erosion problem in Aberystwyth, the causes of which were analysed by So (1974, p. 151):

> The problem in Aberystwyth in general, and at Victoria Terrace in particular, is a direct outcome of human attempts at immediate rewards without visualizing the possible adverse consequences that may follow. The harbour works with the stone pier are bulky enough to aggravate the supply of beach sediments, but not to protect the beaches from the effect of onshore gales

especially from the south-west. The situation is aggravated by the building of the town on a storm beach and its extension to the tide-water level, thus interfering with the natural coastal hinterland across the path of advancing dominant waves . . . To this is added the friable nature of the parent rock which results in beach materials being continuously broken down.

The situation in Bournemouth, where in the nineteenth century the Undercliff Drive was constructed to stop cliff erosion and allow access to the beach, has been described by Carter (1987, p. 47):

However, the cliff was the main source of sand to the beach, and when the cliff erosion was stopped, the remaining beach sand was washed away along the shore, leaving the sea wall badly exposed. So groynes were built, then more groynes, followed by a bigger and 'better' sea wall and the cliff was drained and concreted. In the 1970s, the beach was 'nourished' by pumping sand on to the beach from offshore, thus re-establishing a natural cycle that was broken 150 years before.

Coastal dunes are a landscape type that is highly susceptible to erosion. Indeed, there are relatively few dune systems in England and Wales which are accreting to any great extent. Exceptions include Morfa Harlech and Morfa Dyffryn in North Wales, and Tywyn Point in South Wales. The most frequent cause of erosion is human feet and vehicles (Liddle and Greig-Smith, 1975; Doody, 1985). Together they may kill the dune vegetation, exposing the underlying sand to the action of wind, resulting in a 'blow out' and causing the loss of large areas of vegetation and sand. Timber walkways, control of access routes and vegetation management may help to alleviate the problem. This is important because dunes are not only valuable natural habitats, they also perform an important sea-defence function.

Some of the most profound human modifications of the coastline may lie in the near future. If a climatic warming results from various human modifications of the atmosphere, most notably through the emission of carbon dioxide and other 'greenhouse' gases, then sea-level rise will occur, creating not only accelerated coastal flooding, but accelerated loss of beaches, dunes, salt marshes and cliffs.

REFERENCES

Ackermann, K. J. and Cave, R. 1967: Superficial deposits and structures, including landslip in the Stroud district, Gloucestershire. *Proceedings of the Geologists' Association of London* 78, 567–86.

Adam, P. 1978: Geographical variation in British saltmarsh vegetation. *Journal of Ecology*, 66, 339–66.

Adam, P. 1981: The vegetation of British saltmarshes. *New Phytologist* 88, 143–96.

Addison, K. 1987: Debris flow during intense rainfall in Snowdonia, North Wales: a preliminary survey. *Earth Surface Processes and Landforms*, 12, 561–66.

Agar, R. 1960: Post-glacial erosion of the North Yorkshire coast from the Tees estuary to Ravenscar. *Proceedings Yorkshire Geological Society*, 32, 409–27.

Alcock, M. R. and Morton, A. J. 1985: Nutrient content of throughfall and stemflow in woodland recently established on heathland. *Journal of Ecology*, 73, 625–32.

Allen, J. R. L. 1984: Truncated fossil thermal contraction polygons (Devensian) in the Mercia Mudstone Formation (Trias), Oldbury upon Severn, Gloucestershire. *Proceedings Geologists' Association*, 95 (3), 263–73.

Allen, J. R. L. 1985: *Principles of Physical Sedimentology*, London: Allen and Unwin 272–88.

Allen, J. R. L. 1989: Evolution of salt-marsh cliffs in muddy and sandy systems: a qualitative comparison of British west-coast estuaries. *Earth Surface Processes and Landforms*, 14, 85–92.

Allen, J. R. L. and Rae, J. E. 1987: Late flandrian shoreline oscillations in the Severn Estuary: A geomorphological and stratigraphical reconnaissance. *Philosophical Transactions of the Royal Society London* B, 315, 185–230.

Allen, J. R. L. and Rae, J. E. 1988: Vertical salt-marsh accretion since the Roman period in the Severn Estuary, southwest Britain. *Marine Geology*, 83, 225–235.

Allen, P. M. and Jackson, A. A. 1985: *Geology of the country round Harlech*, British Geological Survey, London: HMSO.

Allen, S. E., Carlisle, A., White, E. J. and Evans, C. C. 1968: The plant nutrient content of rainwater. *Journal of Ecology*, 56, 497–504.

Anderson, E. W. 1972: Terracettes: a suggested classification. *Area*, 4, 17–20.

Anderson, E. W. and Cox, N. J. 1984: The relationship between soil creep rate and certain controlling variables in a catchment in upper Weardale, northern England. In T. P. Burt and D. E. Walling (eds), *Catchment Experiments in Fluvial Geomorphology*, Norwich: Geobooks, 419–30.

Anderson, J. G. C. and Owen, T. R. 1980: *The structure of the British Isles*, 2nd edn, Oxford: Pergamon Press, 251 pp.

Anderson, M. G. and Calver, A. 1977: On the persistence of landscape features formed by a large flood. *Transactions Institute of British Geographers*, 2, 234–54.

Anderson, M. G. and Richards, K. S. 1981: Geomorphological aspects of slopes in mudrocks of the United Kingdom. *Quarterly Journal of Engineering Geology*, 14, 363–72.

Anderton, R., Bridges, P. H., Leader, M. R. and Sellwood, B. W. 1979: *A Dynamic Stratigraphy of the British Isles: A Study in Coastal Evolution*, London: Allen & Unwin, 301 pp.

Andrews, J. T., Bowen, D. Q. and Kidson, C. 1979: Amino acid ratios and the correlation of raised beach deposits in south-west England and Wales. *Nature*, 281, 556–8.

Arber, E. A. N. 1911: *The Coast Scenery of North Devon*.

Arber, M. A. 1946: Dust-storms in the Fenland around Ely. *Geography*, 31, 23–6.

Arkell, W. J. 1947: *The Geology of Oxford*, Oxford: Clarendon Press.

Arthurton, R. S. and Wadge, A. J. 1981: *Geology of the Country around Penrith*, IGS, London: HMSO.

Atkinson, T. C., Briffa, K. R. and Coope, G. R. 1987: Seasonal temperatures in Britain during the past 22,000 years, reconstructed using beetle remains. *Nature*, 325, 587–92.

Atkinson, T. C., Harmon, R. S., Smart, P. L. and Waltham, A. C. 1978: Palaeoclimatic and geomorphic implication of ^{230}Th/^{234}U dates on speleothems from Britain. *Nature*, 272, 24–28.

Atkinson, T. C., Lawson, T. J., Smart, P. L., Harmon, R. S. and Hess, J. W. 1986: New data on speleothem deposition and palaeoclimate in Britain over the last forty thousand years. *Journal of Quarternary Science*, 1, 67–72.

Atkinson, T. C. and Smith, D. I. 1974: Rapid groundwater flow in fissures in the chalk: an example from South Hampshire. *Quarterly Journal of Engineering Geology*, 7, 197–205.

Baker, C. A. and Jones, D. K. C. 1980: Glaciation of the London Basin and its influence on the drainage pattern. In D. K. C. Jones (ed.), *The Shaping of Southern England*, London: Academic Press, pp. 131–75.

Balchin, W. G. V. 1937: The erosion surfaces of north Cornwall. *Geographical Journal*, 90, 52–63.

Balchin, W. G. V. 1952: The erosion surfaces of Exmoor and adjacent areas. *Geographical Journal*, 118, 453–76.

Balescu, S. and Hasgaerts, P. 1984: The Sangatte raised beach and the age of the opening of the strait of Dover. *Geologie en Mijubouw*, 63, 355–62.

Ball, D. F. 1966: Late-glacial scree in Wales. *Biulteyn Peryglacjalny*, 15, 151–63.

Ball, D. F. and Goodier, R. 1970: Morphology and distribution of features resulting from frost-action in Snowdonia. *Field Studies*, 3, 193–218.

Ballantyne, C. K. 1987: The present-day periglaciation of upland Britain. In J. Boardman (ed.), *Periglacial Processes and Landforms in Britain and Ireland*, Cambridge: Cambridge University Press, 113–26.

Ballantyne, C. K. and Kirkbride, M. P. 1986: The characteristics and significance of some late glacial protalus ramparts in upland Britain. *Earth Surface Processes and Landforms*, 11, 659–71.

Ballantyne, C. D. and Whittington, G. 1987: Niveo-aeolian sand deposits on An Teallach, western Ross, Scotland. *Transactions of the Royal Society of Edinburgh: Earth Sciences*, 78, 51–63.

Balson, P. S. and Humphreys, B. 1986: The nature and origin of fissures. The East Anglian Coralline and Red Crags. *Journal of Quaternary Science*, 1, 13–19.

Banham, P. H. 1975: Glacitectonic structures: a general discussion with particular reference to the contorted drift of Norfolk. In A. E. Wright and F. Moseley, (eds), *Ice Ages: Ancient and Modern*, Liverpool: Seel House, 69–94.

Barker, R. D. and Harker, D. 1984: The location of the Stour buried tunnel-valley using geophysical techniques. *Quarterly Journal of Engineering Geology*, 7, 103–15.

Barnes, F. A. 1963: Peat erosion in the Southern Pennines: problems of interpretation. *East Midland Geographer*, 3 (4), 216–22.

Barron, E. J. 1985: Explanations of the Tertiary global cooling trend. *Palaeogeograhy, Palaeoclimatology, Palaeoecology*, 50, 45–61.

Barrow, G. 1908: The high level platforms of Bodmin Moor and their relation to the deposits of stream-tin and wolfram. *Quarterly Journal of the Geological Society of London*, 64, 384–400.

Barton, M. E. and Coles, B. J. 1984: The characteristics and rates of the various slope degradation processes in the Barton Clay Cliffs of Hampshire. *Quarterly Journal Engineering Geology*, 17, 117–36.

Barton, M. E., Coles, B. J. and Tiller, G. R. 1983: A statistical study of the cliff top slumps in part of the Christchurch Bay coastal cliffs. *Earth Surface Processes and Landforms*, 8, 409–22.

Battarbee, R. W., Appleby, P. G., Odell, K. and Flower, R. J. 1985: [210]Pb dating of Scottish lake sediments, afforestation and accelerated soil erosion. *Earth Surface Processes and Landforms*, 10, 137–42.

Battiau-Queney, Y. 1981: Les effets géomorphologiques des glaciations Quaternaires au pays-de-Galles (Grande Bretagne). *Revue de Géomorphologie Dynamique*, 30, 63–73.

Battiau-Queney, Y. 1984: The pre-glacial evolution of Wales. *Earth Surface Processes and Landforms*, 9, 229–52.

Bazley, R. A. and Bristow, C. R. 1972: *Geology of the Country around Royal Tunbridge Wells*, London: HMSO.

Beckinsale, R. P. 1970: Physical problems of Cotswold rivers and valleys. *Proceedings Cotteswold Naturalists' Field Club*, 35, 194–205.

Belbin, S. 1985: Long-term landform development in North-West England: the application of the planation concept. In R. H. Johnson (ed.), *The Geomorphology of North-West England*, Manchester: Manchester University Press, 37–58.

Bell, F. G. 1983: *Engineering Properties of Soils and Rocks*. London: Butterworth.

Bell, M. L. 1982: The effect of land-use and climate on valley sedimentation. In A. F. Harding (ed.), *Climate Change in later Prehistory*, Edinburgh: Edinburgh University Press, 127–42.

Bennett, F. J. 1908: Solution-subsidence valleys and swallow-holes within the Hythe Beds area of West Malling and Maidstone. *Geographical Journal*, 32, 277–88, 291–3.

Bevan, K., Lawson A. and McDonald, A. 1978: A landslip/debris flow in Bilsdale, North York Moors, September 1976. *Earth Surface Processes*, 3, 407–19.

Bhadresa, R. 1977: Food preferences of rabbits *Oryctolagus curiculus* L. at Holkham sand dunes, Norfolk. *Journal of Applied Ecology*, 14, 287–91.

Bird, E. C. F. 1984: *Coasts: an introduction to coastal geomorphology*, 3rd edn, Oxford: Basil Blackwell.

Bird, E. C. F. and May, V. J. 1976: *Shoreline Changes in the British Isles during the Past Century*. Bournemouth College of Technology, Division of Geography: Bournemouth.

Bleasdale, A. 1970: The rainfall of 14th and 15th September 1968 in comparison with previous exceptional rainfall in the United Kingdom. *Journal Institute of Water Engineers*, 24, 181–9.

Blenkinsop, T. H., Long, R. E., Kusznir, N. J. and Smith, M. J. 1986: Seismicity and tectonics in Wales. *Journal of the Geological Society, London*, 143, 327–34.

Bloom, A. L. 1971: Glacial eustatic and isostatic controls of sea-level since the Last Glaciation. In K. K. Turekian (ed.), *The Late Cainozoic Glacial Ages*, New Haven: Yale University Press, 355–79.

Boardman, J. 1978: Grèzes litées near Keswick, Cumbria. *Biulteyn Peryglacjalny*, 27, 23–34.

Boardman, J. C. (ed.) 1985: *Field guide to the periglacial landforms of Northern England*. Cambridge: Quaternary Research Association.

Boardman, J. 1988: *Classic landforms of the Lake District*, Sheffield: Geographical Association.

Boardman, J. and Robinson, D. A. 1985: Soil erosion, climatic vagary and agricultural change on the Downs around Lewes and Brighton, Autumn 1982. *Applied Geography*, 5, 243–58.

Boulton, G. S. 1977: In R. G. West (ed.), *INQUA Guidebook to East Anglia*.

Boulton, G. S., Cox, F., Hart, J. and Thornton, M. 1984: The glacial geology of Norfolk. *Bulletin, Geological Society of Norfolk*, 34, 103–22.

Boulton, G. S., Jones, A. S., Clayton, K. M. and Kenning, M. J. 1977: A British ice-sheet model and patterns of glacial erosion and deposition in Britain. In F. W. Shotton, (ed.), *British Quaternary Studies – Recent Advances*, Oxford: Clarendon Press, 231–46.

Bowen, D. Q. 1973: The Pleistocene succession of the Irish Sea. *Proceedings Geologists' Association*, 84, 249–72.

Bowen, D. Q. 1977: *INQUA guidebook for excursions A8 and C8, Wales and the Cheshire–Shropshire lowland*.

Bowen, D. Q. 1978: *Quaternary Geology*. Oxford: Pergamon.

Bowen, D. Q., Rose, J., McCane, A. M. and Sutherland, D. G. 1986: Correlation of Quaternary glaciation in England, Ireland, Scotland and · Wales. *Quaternary Science Reviews*, 5, 299–340.

Bower, M. M. 1961: The distribution of erosion in blanket peat bogs in the Pennines. *Transactions, Institute of British Geographers*, 29, 17–30.

Bridgland, D. R. 1988: The Pleistocene fluvial stratigraphy and palaeo-geography of Essex. *Proceedings Geologists' Association*, 99, 291–314.

Briggs, D. J. 1975: Origin, depositional environment and age of the Cheltenham sand and gravel and related deposits. *Proceedings Geologists' Association*, 86, 333–48.

Briggs, D. J. and Courtney, F. M. 1972: Ridge-and-trough topography in the North Cotswolds. *Proceedings Cotteswold Naturalists' Field Club*, 36, 94–103.

Briggs, D. J. and Gilbertson, D. D. 1980: Quaternary process and environments in the Upper Thames Valley. *Transactions, Institute of British Geographers*, 5, 53–65.

Bristow, C. R. and Bazley, R. A. 1972: *Geology of the Country around Royal Tunbridge Wells*. London: HMSO.

Bromhead, E. N. 1978: Large landslides in London Clay at Herne Bay, Kent. *Quarterly Journal of Engineering Geology*, 11, 291–304.

Bromhead, E. N. 1979: Factors affecting the transition between the various types of mass movement in coastal cliffs consisting largely of overconsolidated clay with special reference to Southern England. *Quarterly Journal of Engineering Geology*, 12, 291–300.

Brookes, A. 1987: River channel adjustments downstream from

channelization works in England and Wales. *Earth Surface Processes and Landforms*, 12, 337–51.

Brookes, A., 1988: *Channelized Rivers: Perspectives for Environmental Management*. Chichester: John Wiley.

Brookes, A., Gregory, K. J. and Dawson, F. H. 1983: An assessment of river channelization in England and Wales. *Science of the Total Environment*, 27, 97–112.

Brown, E. H. 1960a: The building of southern Britain. *Zeitschrift für Geomorphologie*, 4 (3–4), 267–74.

Brown, E. H. 1960b: *The Relief and Drainage of Wales*. Cardiff: University of Wales Press, 186 pp.

Brown, E. H. 1961: Britain and Appalachia: a study in the correlation and dating of planation surfaces. *Transactions, Institute of British Geographers*, 29, 91–100.

Brown, E. H. 1969: Jointing, aspect, and the orientation of scarp face dry valleys, near Ivinghoe, Buckinghamshire. *Transactions, Institute of British Geographers*, 48, 61–73.

Brown, E. H. 1979: The shape of Britain. *Transactions, Institute of British Geographers*, NS 4, 449–62.

Brown, I. W., Wathern, P., Roberts, D. A. and Young, S. N. 1985: Monitoring sand dune erosion on the Clwyd coast, North Wales. *Landscape Research*, 10, 14–17.

Brunsden, D. 1968: Dartmoor. *British Landscapes Through Maps*, 13.

Brunsden, D. 1979: Mass movements. In C. Embleton, and J. Thornes (eds), *Process in Geomorphology*, London: Edward Arnold, 130–86.

Brunsden, D., Doornkamp, J. C., Green, C. P. and Jones, D. K. C. 1976: Tertiary and Cretaceous sediments in solution pipes in the Devonian Limestone of South Devon, England. *Geological Magazine*, 113, 441–7.

Brunsden, D. and Goudie, A. S. 1981: *Classic Landforms of the Dorset Coast*. Sheffield: Geographical Association.

Brunsden, D. and Jones, D. K. C. 1976: The evolution of landslide slopes in Dorset. *Philosophical Transactions, Royal Society London* A, 283, 605–31.

Brunsden, D., Kidson, C., Orme, A. R. and Waters, R. S. 1964: The denudation chronology of parts of south-western England. *Field Studies*, 2, 115–32.

Bryant, I. D. 1983: The utilization of Arctic river analogue studies in the interpretation of periglacial river sediments from southern Britain. In K. J. Gregory (ed.), *Background to Palaeohydrology*, Chichester: John Wiley, pp. 413–31.

Bryant, R. H. and Carpenter, C. 1987: Ramparted ground ice depressions in Britain and Ireland. In J. Boardman (ed.), *Periglacial Processes and Landforms in Britain and Ireland*, Cambridge: Cambridge University Press, 183–90.

Buchardt, B. 1978: Oxygen isotope palaeotemperatures from the Tertiary period in the North Sea area. *Nature*, 275, 121–3.

Buckland, W. 1840/41: On the evidences of glaciers in Scotland and the north of England. *Proceedings Geological Society*, 3, 332–7, 345–8.

Büdel, J. 1982: *Climatic Geomorphology*. Princeton: Princeton University Press.

Bull, P. A. 1980: The antiquity of caves and dolines in the British Isles. *Zeitschrift für Geomorphologie Supplementband*, 36, 217–32.

Bunting, B. T. 1964: Slope development and soil formation on some British Sandstones. *Geographical Journal*, 130, 73–9.

Burgess, I. C. and Holliday, D. W. 1979: *Geology of the country around Brough-under-Stainmore*, London: IGS/HMSO.

Burrin, P. J. 1981: Loess in the Weald. *Proceedings Geologists' Association*, 92, 87–92.

Burrin, P. J. 1985: Holocene alluviation in southeast England and some implications for palaeohydrological studies. *Earth Surface Processes and Landforms*, 10, 257–71.

Burrin, P. J. and Scaife, R. G. 1984: Aspects of Holocene valley sedimentation and floodplain development in southern England. *Proceedings Geologists' Association*, 95, 81–96.

Burt, T. P., Butcher, D. P., Coles, N. and Thomas, A. D. 1983: The natural history of Slapton Ley Nature Reserve XV: hydrological processes in the Slapton Wood catchment. *Field Studies*, 5, 731–52.

Burt, T. P., Donohoe, M. A. and Vann, A. R. 1983: The effect of forestry drainage operations on upland sediment yields: the results of a storm-based study. *Earth Surface Processes and Landforms*, 8, 339–46.

Burton, R. G. O. 1976: Possible thermokarst features in Cambridgeshire. *East Midland Geographer*, 6 (5), 230–40.

Bury, H. 1910: On the denudation of the western end of the Weald. *Quarterly Journal Geological Society of London*, 66, 640–92.

Butler, P. B. 1983: Landsliding and other large scale mass movements on the escarpment of the Cotswold Hills. BA dissertation, unpublished, School of Geography, University of Oxford, 146 pp.

Buuman, P. 1980: Palaeosols in the Reading Beds (Palaeocene) of Alum Bay, Isle of Wight, UK. *Sedimentology*, 27, 593–606.

Caine, T. N. 1972: The distribution of sorted patterned ground in the English Lake District. *Revue de Géomorphologie dynamique*, 21, 49–56.

Cameron, T. D. J., Stoker, M. S. and Long, D. 1987: The history of Quaternary sedimentation in the UK sector of the North Sea Basin. *Journal of the Geological Society, London*, 144, 43–58.

Carling, P. A. 1983: Particulate dynamics, dissolved and total load, in two small basins, northern Pennines, UK. *Hydrological Sciences Journal*, 28, 355–75.

Carling, P. A. 1986: Peat slides in Teesdale and Weardale, Northern Pennines, July 1983: Description and failure mechanisms. *Earth Surface Processes and Landforms*, 11, 193–206.

Carling, P. A. 1987: A terminal debris-flow lobe in the northern Pennines,

United Kingdom. *Transactions Royal Society of Edinburgh, Earth Sciences* 78, 169–76.

Carr, A. 1965: Shingle spit and river mouth: short-term dynamics. *Transactions, Institute of British Geographers*, 36, 117–130.

Carr, A. P. and Blackley, M. W. 1973: Investigations bearing on the age and development of Chesil Beach, Dorset and the associated area. *Transactions, Institute of British Geographers*, 58, 98–111.

Carr, A. P. and Blackley, W. L. 1986: Seasonal changes in surface level of a salt marsh creek. *Earth Surface Processes and Landforms*, 11, 427–39.

Carr, A. P. and Graff, J. 1982: The tidal immersion factor and shore platform development. *Transactions, Institute of British Geographers*, NS 7, 240–5.

Carson, M. A. and Kirkby, M. J. 1972: *Hillslope Form and Process*. Cambridge: Cambridge University Press.

Carson, M. A. and Petley, D. J. 1970: The existence of threshold hillslopes in the denudation of the landscape. *Transactions, Institute of British Geographers*, 49, 71–95.

Carter, W. 1987: Guarding the Frontier. *Landscape*, 1 (2), 45–48.

Castleden, R. 1976: The floodplain gravels of the River Nene. *Mercian Geologist*, 6 (1), 33–47.

Castleden, R. 1977: Periglacial sediments in central and southern England. *Catena*, 4, 111–21.

Catt, J. A. 1977: Loess and Coversands. In F. W. Shotton (ed.), *British Quaternary Studies: Recent Advances*, Oxford: Clarendon Press, 221–9.

Catt, J. A. 1979: Distribution of loess in Britain. *Proceedings Geologists' Association*, 90, 93–5.

Catt, J. A. 1986a: *Soils and Quaternary Geology. A Handbook for Field Scientists*. Oxford: Clarendon Press.

Catt, J. A. 1986b: The nature, origin and geomorphological significance of clay-with-flints. In G. de G. Sieveking and M. B. Hart (eds), *The Scientific Study of Flint and Chert*, Cambridge, Cambridge University Press, 151–9.

Catt, J. A., Corbett, W. M., Hodge, C. A. H., Madgett, P. A., Tayler, W. and Weir, A. H. 1971: Loess in the soils of north Norfolk. *Journal of Soil Science*, 22, 444–52.

Catt, J. A., Green, M. and Arnold, N. J. 1982: Naleds in a Wessex downland valley. *Proceedings of the Dorset Natural History and Archaeological Society*, 102, 68–75.

Catt, J. A. and Staines, S. J. 1982: Loess in Cornwall. *Proceedings Ussher Society*, 5, 368–75.

Catt, J. A., Weir, A. H. and Madgett, P. A. 1974: The loess of eastern Yorkshire and Lincolnshire. *Proceedings Yorkshire Geological Society*, 40, 23–39.

Chambers, F. M., Dresser, P. Q. and Smith, A. G. 1979: Radiocarbon dating evidence on the impact of atmospheric pollution on upland peats. *Nature*, 282, 829–31.

Chambers, W. 1983: Denudation rates in the River Burry catchment, Gower, Glamorgan. *Cave Science*, 10, 180–7.

Chandler, R. H. 1909: On some dry chalk valley features. *Geological Magazine*, 6, 538–9.

Chandler, R. J. 1970: The degradation of Lias Clay slopes in an area of the East Midlands. *Quarterly Journal of Engineering Geology*, 2, 161–81.

Chandler, R. J. 1976: The history and stability of two Lias clay slopes in the Upper Gwash Valley, Rutland. *Philosophical Transactions Royal Society of London* A, 283, 463–91.

Chandler, R. J., Kellaway, G. A., Skempton, A. W. and Wyatt, R. J. 1976: Valley slope sections in Jurassic strata near Bath, Somerset. *Philosophical Transactions Royal Society of London* A, 285, 527–56.

Chandler, R. J. and Poole, M. J. 1971: Creep movement on low gradient clay slopes since the Late Glacial. *Nature*, 229, 399–400.

Charman, K., Fatt, W. and Penny, S. 1986: *Saltmarsh Survey of Great Britain: Bibliography.* Peterborough: Nature Conservancy Council.

Cheetham, G. H. 1980: Late Quaternary palaeohydrology: The Kennet Valley case-study. in D. K. C. Jones (ed.), *The Shaping of Southern England*, London: Academic Press, 203–23.

Christopher, N. S. J., Beck, J. S. and Mellors, P. T. 1977: Hydrology-water in the limestone. In T. D. Ford (ed.), *Limestone and Caves of the Peak District*, Norwich: Geobooks, 185–229.

Clapperton, C. M. 1971: The location and origin of glacial meltwater phenomena in the east Cheviot Hills. *Proceedings Yorkshire Geological Society*, 38, 361–80.

Clark, M. J. 1965: The form of chalk slopes. *University of Southampton, Dept. of Geography, Research Series*, 2, 3–4.

Clark, M. R. and Dixon, A. G. 1981: The Pleistocene braided river deposits in the Blackwater valley area of Berkshire and Hampshire, England. *Proceedings Geologists' Association*, 92 (3), 139–57.

Clarke, W. G. 1925: *In Breckland Wilds.* London: Scott, 298 pp.

Clayton, K. M. 1953: The denudation chronology of part of the Middle Trent basin. *Transactions, Institute of British Geographers*, 19, 25–36.

Clayton, K. M. 1977: River Terraces. In F. W. Shotton (ed.), *British Quaternary studies recent advances*, Oxford: Clarendon Press, pp. 153–67.

Clayton, K. M. 1979: *The Midlands and Southern Pennines.* In A. Straw and K. Clayton, *Eastern and Central England*, London: Methuen, 143–247.

Clayton, K. M. 1981: Explanatory description of the landforms of the Malham area. *Field Studies*, 5, 389–423.

Clayton, K. M. 1989: Sediment input from the Norfolk cliffs, Eastern England – a century of coast protection and its effect. *Journal of Coastal Research*, 5, 433–42.

Coles, B. P. L. and Funnell, B. M. 1981: Holocene palaeoenvironments of Broadland, England. *Special Publs. Int. Ass. Sediment*, 5, 123–31.

Collinson, M. E., Fowler, K. and Boulter, M. C. 1981: Floristic changes indicate a cooling climate in the Eocene of Southern England. *Nature*, 291, 315–17.

Collinson, M. E. and Hooker, J. J. 1987: Vegetational and mammalian formal changes in the early Tertiary of southern England. In E. M. Friis, W. G. Chaloner and P. R. Crane (eds), *The Origin of Angiosperms and their Biological Consequences*, Cambridge: Cambridge University Press, 259–304.

Cominetti, M. E. 1986: Problems of water quality associated with the mining and extractive industries of South west England – some case histories. In J. F. de L. G. Solbé (ed.), *Effects of Land Use on Fresh Waters*, Chichester: Ellis Horwood, 147–61.

Common, R. 1954: The geomorphology of the east Cheviot area. *Scottish Geographical Magazine*, 70, 124–38.

Conway, B. W. 1979: The contribution made to cliff instability by head deposits in the west Dorset coastal area. *Quarterly Journal of Engineering Geology*, 12, 267–75.

Conway, V. M. 1954: Stratigraphy and pollen analysis of southern Pennine blanket peats. *Journal of Ecology*, 42, 117–47.

Coones, P. and Patten, J. H. C. 1986: *The Landscape of England and Wales*. Harmondsworth: Penguin Books.

Cooper, R. G. 1980: A sequence of landsliding mechanisms in the Hambleton Hills, northern England, illustrated by features at Peak Scar, Hawnby. *Geografiska Annaler*, 62A, 149–56.

Coque-Delhuille, B. 1987: *Les massifs du sud-ouest Anglais et sa bordeure sédimentaire: etude géomorphologique*. Paris. Thése Etat, Université de Paris I – Panthéon, Sorbonne.

Coque-Delhuille, B. and Veyret, Y. 1989: Limite d'englacement et évolution periglaciaire des Iles Scilly: L'interêt des arènes in situ et remaniées. *Zeitschrift für Geomorphologie Supplementband*, 72, 79–96.

Corbel, J. 1957: *Les karsts du Nord-Ouest de l'Europe et de quelques régions de comparaison*. Lyon: Institut des Etudes Rhodaniennes de L'Université de Lyon.

Cox, F. C. 1985: The tunnel-valleys of Norfolk, East Anglia. *Proceedings Geologists' Association*, 96, 357–69.

Coxon, P. 1978: The first record of a fossil naled in Britain. *Quaternary Newsletter*, 24, 9–11.

Crampton, C. B. and Taylor, J. A. 1967: Solifluction terraces in south Wales. *Biulteyn peryglacjalny*, 16, 15–36.

Creber, G. T. and Chaloner, W. G. 1985: Tree growth in the Mesozoic and early Tertiary and the reconstruction of palaeoclimates. *Palaeogeography, Palaeoclimatology, Palaeoecology*, 52, 35–60.

Cripps, J. C. and Taylor, R. K. 1986: Engineering characteristics of British over-consolidated clays and mudrocks, II. Tertiary deposits. *Engineering Geology*, 22, 349–76.

Cripps, J. C. and Taylor, R. K. 1987: Engineering characteristics of

British over-consolidated clays and mudrocks, II. Mesozoic deposits. *Engineering Geology*, 23, 213–53.

Crisp, D. T. 1966: Input and output of minerals for an area of Pennine moorland: the importance of precipitation, drainage, peat erosion and animals. *Journal of Applied Ecology*, 3, 327–48.

Crisp, D. T., Rawes, M. and Welch, D. 1964: A Pennine peat slide. *Geographical Journal*, 130, 519–24.

Crisp, D. T. and Robson, S. 1979: Some effects of discharge upon the transport of animals and peat in a North Pennine headstream. *Journal of Applied Ecology*, 16, 721–36.

Cryer, R. 1980: The chemical quality of some pipeflow waters in upland mid-Wales and its implications. *Cambria*, 6, 28–46.

Cunliffe, B. W. 1980: The evolution of Romney Marsh: a preliminary statement. In Thompson, F. H. (ed.), *Archaeology and Coastal Change*, London: Society of Antiquaries, 37–55.

Cunningham, F. F. 1965: Tor Theories in the light of South Pennine evidence. *East Midlands Geographer*, 3, 424–33.

Curtis, L. F., Courtney, F. M. and Trudgill, S. 1976: *Soils in the British Isles*. London: Longman.

Dalby, D. H. 1970: The salt marshes of Milford Haven, Pembrokeshire. *Field Studies*, 3, 297–330.

Daley, B. 1972: Some problems concerning the Early Tertiary climate of southern Britain. *Palaeogeography, Palaeoclimatology, Palaeoecology*, 11, 177–190.

Dalton, R. T. and Fox, H. R. 1986: Dry valleys in the Denton area of south Derbyshire. *Mercian Geologist*, 10 (2), 77–90.

Dalzell, D. and Durrance, E. M. 1980: The evolution of the Valley of Rocks, North Devon. *Transactions, Institute of British Geographers*, NS 5, 66–79.

Darwin, C. 1842: Notes on the effects by the ancient glaciers of Caernarvonshire, etc. *Philosophical Magazine*, 21, 180–8.

Darwin, C. 1882: *Vegetable mould and earth-worms*. London: Murray, 328 pp.

Davies, K. H. and Keen D. H. 1985: The age of Pleistocene marine deposits at Portland, Dorset. *Proceedings of the Geologists' Association*, 96, 217–25.

Davis, W. M. 1895: The development of certain English rivers. *Geographical Journal*, V, 127–46.

Davis, W. M. 1909: Glacial erosion in north Wales. *Quarterly Journal Geological Society of London*, 65, 281–350.

Davison, C. 1924: *A History of British Earthquakes*. Cambridge: Cambridge University Press, 416 pp.

Dawson, M. R. and Gardiner, V. 1987. River terraces: the general model and a palaeohydrological and sedimentological interpretation of the terraces of the Lower Severn. In K. J. Gregory, J. Lewin and J. B. Thornes (eds), *Palaeohydrology in Practice*, London: John Wiley, 269–305.

Deadman, A. 1984: Recent history of Spartina in north west England and in North Wales and its possible future development. In P. Doody (ed.) *Spartina anglica in Great Britain*, Shrewsbury: Nature Conservancy Council, 22–4.

De Boer, G. 1964: Spurn Head: its history and evolution. *Transactions, Institute of British Geographers*, 34, 71–89.

Defoe, D. 1971: *A Tour Through the Whole Island of Great Britain*. Harmondsworth: Penguin Books, 730 pp.

De Freitas, M. H. 1972: Some examples of cliff failure in south-west England. *Proceedings Ussher Society*, 2, 388–97.

de Gans, W. 1988: Pingo scars and their identification. In M. J. Clark (ed.), *Advances in Periglacial Geomorphology*, Chichester: John Wiley, pp. 299–322.

De La Beche, H. T. 1839: *Report on the Geology of Cornwall, Devon and west Somerset*. Memoir Geological Survey of Great Britain, 648 pp.

Denness, B., Conway, B. W., McCann, D. M. and Grainger, P. 1975: Investigation of a coastal landslip at Charmouth, Dorset. *Quarterly Journal of Engineering Geology*, 8, 119–40.

Derbyshire, E. 1979: Recent movements on the cliff at St. Mary's Bay, Brixham, Devon. *Geographical Journal*, 145, 88.

Destombes, J.-P., Shephard-Thorn, E. R. and Redding, J. H. 1975: A buried valley system in The Strait of Dover. *Philosophical Transactions Royal Society of London* A, 279, 243–56.

Devoy, R. J. N. 1982: Analysis of geological evidence for Holocene sea-level movements in south-east England. *Proceedings Geologists' Association* 93, 65–90.

Devoy, R. J. 1985: The problem of a Late Quaternary landbridge between Britain and Ireland. *Quaternary Science Reviews* 4, 43–58.

Dewey, J. F. 1982: Plate tectonics and the evolution of the British Isles. *Journal of the Geological Society of London*, 139, 371–412.

Dewey, J. F. and Windley, B. F. 1988: Palaeocene-Oligocene tectonics of NW Europe. *Geological Society Special Publication*, 39, 25–31.

Dewolf, Y. and Mainguet, M. 1976: Une hypothèse éolienne et tectonique sur l'alignement et l'orientation des buttes Tertiaires du Bassin de Paris. *Revue de Géologie Dynamique*, 18, 415–26.

Dines, H. G., Hollingworth, S. E., Edwards, W., Beecham, S. and Welch, F. B. A. 1940: The mapping of head deposits. *Geological Magazine*, 77, 198–226.

Diver, C. 1933: The physiography of South Haven peninsula, Studland Heath, Dorset. *Geographical Journal*, 81, 404–27.

d'Olier, B. 1975: Some aspects of the Late Pleistocene-Holocene drainage of the river Thames in the eastern part of the London Basin. *Philosphical Transactions Royal Society of London* A, 279, 269–77.

Donn, W. L., Farrand, W. R. and Ewing, M. 1962: Pleistocene ice volumes and sea-level changes. *Journal of Geology*, 70, 206–14.

Donovan, D. T. and Stride, A. M. 1975: Three drowned coast-lines of probable late tertiary age around Devon and Cornwall. *Marine Geology*, 19, M35–40.

Doody, P. (ed.) 1984: *Spartina anglica in Great Britain*. Shrewsbury: Nature Conservancy Council, 72 pp.

Doody, P. (ed.) 1985: *Sand Dunes and their Management*. Peterborough: Nature Conservancy Council, 262 pp.

Doody, P. and Barnett, B. (eds) 1987: *The Wash and its Environment*. Peterborough: Nature Conservancy Council, 208 pp.

Doornkamp, J. C. 1974: Tropical weathering and the ultra-microscopic characteristics of regolith quartz on Dartmoor. *Geografiska Annaler* 56A, 73–82.

Douglas, I. 1987: Plate tectonics, palaeoenvironments and limestone geomorphology in west-central Britain. *Earth Surface Processes and Landforms*, 12, 481–95.

Douglas, T. D. and Harrison, S. 1987: Late Devensian periglacial slope deposits in the Cheviot Hills. In J. Boardman (ed.) *Periglacial Processes and Landforms in Britain and Ireland*, Cambridge: Cambridge University Press, 237–44.

Driscoll, E. M. 1958: The denudation chronology of the Vale of Glamorgan. *Transactions of the Institute of British Geographers*, 25, 45–57.

Dunning, F. W., Mercer, I. R., Owen, M. P., Roberts, R. H. and Lambert, J. L. M. 1978: *Britain before Man*. London: HMSO.

Durrance, E. R. 1969: The buried channels of the Exe. *Geological Magazine*, 106 (2), 174–89.

Dury, G. H. 1954: Weather, climate and river erosion in the Ice Age. *Science News*, 33, 65–88.

Dury, G. H. 1965: Theoretical implications of underfit streams. *United States Geological Survey Professional Paper*, 452-C, 53.

Dury, G. H. 1972: A partial definition of the term *pediment* with field tests in humid-climate areas of southern England. *Transactions, Institute of British Geographers*, 57, 139–52.

Dury, G. H. 1986: *The Face of the Earth*. London: Unwin Hyman.

Early, K. R. and Jordan, P. G. 1985: Some landslipping encountered during construction of the A40 near Monmouth. *Quarterly Journal of Engineering Geology, London*, 18, 207–24.

Eastwood, T., Dixon, E. E. L., Hollingworth, S. E. and Smith, B. 1931: *The Geology of the Whitehaven and Workington District*. London: Memoir Geological Survey of England and Wales, 304 pp.

Eden, M. J. and Green, C. P. 1971: Some aspects of granite weathering and tor formation on Dartmoor, England. *Geografiska Annaler*, 53A, 92–9.

Edmonds, C. N. 1983: Towards the prediction of subsidence risk upon the Chalk outcrop. *Quarterly Journal of Engineering Geology*, 16, 261–6.

Edmunds, F. H. 1944: Small holes and springs in the Chalk of the Mole valley. *London Naturalist* (for 1943), 2–7.

Edwards, A. M. 1973: Dissolved load and tentative solute budgets of some Norfolk catchments. *Journal of Hydrology*, 18, 210–17.

Ellis-Gruffydd, I. D. 1977: Late Devensian glaciation in the Upper Usk basin. *Cambria*, 4, 46–55.

Embleton, C. (ed.) 1984: *Geomorphology of Europe*. London: Macmillan.

Embleton, C. and King C. A. M. 1967: *Glacial and Periglacial Geomorphology*. London: Edward Arnold.

Evans, G. 1979: Quaternary transgressions and regressions. *Journal of the Geological Society*, 136, 125–32.

Evans, J. G. 1966: Late-glacial and post-glacial sub-aerial deposits at Pitstone, Buckinghamshire. *Proceedings Geologists' Association*, 77, 347–64.

Evans, R. 1976: Observations on a stripe pattern. *Biulteyn Peryglacjalny*, 25, 9–22.

Evans, R. and Cook, S. 1986: Soil erosion in Britain. *Seesoil (The Journal of the South East England Soils Discussion Group)*, 3, 28–58.

Everard, C. E. 1957: Erosion platforms on the borders of the Hampshire Basin. *Transactions, Institute of British Geographers*, 22, 33–46.

Eyles, N., Eyles, C. H. and McCabe, A. M. 1989: Sedimentation in an ice-contact subaqueous setting: the Mid-Pleistocene 'North Sea Drifts' of Norfolk, UK. *Quaternary Science Review*, 8, 57–64.

Fagg, C. C. 1939: Physiographical evolution in The North Croydon survey areas and its effect upon vegetation. *Proceedings and Transactions Croydon Natural History and Scientific Society*, 2, 29–60.

Fearnsides, W. G. 1910: North and Central Wales. In H. W. Monckton and R. Harries (ed.), *Geology in the Field*, pp. 786–825, London: Stanford.

Ferguson, R. 1981: Channel form and channel changes. In J. Lewin (ed.), *British Rivers*, London: Allen & Unwin, 90–125.

Ferry, B. and Waters, S. 1985: *Dungeness: Ecology and Conservation*. Peterborough: Nature Conservancy Council, 144 pp.

Fillenham, L. F. 1963: Holme Fen Post. *Geographical Journal*, 129, 502–3.

Finlayson, B. L. 1976: *Measurements of geomorphic processes in a small drainage basin*, unpublished PhD Thesis, University of Bristol.

Finlayson, B. L. 1978: Suspended solids transport in a small experimental catchment. *Zeitschrift für Geomorphologie*, 22, 192–210.

Finlayson, B. L. 1981: Field measurement of soil creep. *Earth Surface Processes and Landforms*, 6, 35–48.

Finlayson, B. L. 1985: Soil creep: a formidable fossil of misconception. In K. S. Richards, R. R. Arnett and S. Ellis (eds) *Geomorphology and soils*. London: Allen & Unwin, 141–58.

Firth, F. M. 1971: *The dune complex at Camber Sands, Sussex*, unpublished PhD Thesis, University of London.

Fisher, O. 1866: On the warp, its age and probable connection with the last geological events. *Quarterly Journal of Geological Society of London*, 22, 553–65.

Fishwick, A. 1977: The Conway Basin. *Cambria*, 4, 56–64.

Flenley, J. R. 1987: The meres of Holderness. In S. Ellis (ed.), *East Yorkshire – Field Guide*, Cambridge: Quaternary Research Association, 73–81.

Ford, D. C. and Stanton, W. I. 1968: The geomorphology of the south-central Mendip Hills. *Proceedings Geologists' Association*, 79, 401–27.

Ford, T. D. 1967: Deep weathering glaciation and tor formation in Charnwood Forest, Leicestershire. *Mercian Geologist*, 2, 3–14.

Ford, T. D. 1984: Palaeokarsts in Britain. *Cave Science*, 11, 246–64.

Ford, T. D. 1986: The evolution of the Castleton Cave Systems, Derbyshire. *Cave Science*, 13, 131–48.

Ford, T. D. 1986: The evolution of the Castleton Cave Systems and related features, Derbyshire. *Mercian Geologist*, 10, 91–114.

Ford, T. D. and Burek, C. V. 1976: Anomalous limestone gorges in Derbyshire. *Mercian Geologist*, 6, 59–66.

Ford, T. D., Gascoyne, M. and Beck, J. S. 1983: Speleothem dates and Pleistocene chronology in the Peak District of Derbyshire. *Cave Science*, 10, 103–15.

Forster, A. and Northmore, K. J. 1985: Landslide distribution in the South Wales coalfield. In C. S. Morgan (ed.), *Landslides in the South Wales Coalfield*, Pontypridd: Polytechnic of Wales, 29–36.

Foster, C. Le N. and Topley, W. 1865: On the superficial deposits of the valley of the Medway, with remarks on the denudation of the Weald. *Quarterly Journal Geological Society of London*, 21, 443–74.

Foster, H. D. 1970: Establishing the age and geomorphological significance of sorted-stripes in the Rhinog Mountains, North Wales. *Geografiska Annaler* A, 52, 96–102.

Foster, I. D. L. 1979: Chemistry of bulk precipitation, throughfall, soil water and stream water in a small catchment in Devon, England. *Catena*, 6, 145–55.

Foster, I. D. L. 1980: Chemical yields in runoff, and denudation in a small arable catchment, East Devon, England. *Journal of Hydrology*, 47, 349–68.

Foster, I. D. L. and Grieve, I. C. 1984: Some implications of small catchment solute studies for geomorphological research. In T. P. Burt and D. Walling (eds), *Catchment Experiments in Fluvial Geomorphology*, Norwich: Geo Books, ch. 22.

Foster, I. D. L. and Grieve, I. C. 1984: Some implications of small catchment solute studies for geomorphological research. In T. P. Burt and D. Walling (eds), *Catchment Experiments in fluvial Geomorphology*, Norwich: Geo Books, ch. 22.

Foster, S.. W. 1987: The dry drainage system on the Northern Yorkshire Wolds. In S. Ellis (ed.) *East Yorkshire – Field Guide*, Cambridge: Quarterly Research Association, 36–7.

Freeman, T. W., Rogers, H. B. and Kinvig, R. H. 1966: *Lancashire, Cheshire and the Isle of Man*, London: Nelson, 308 pp.

French, H. M. 1972: Asymmetrical Slope development in the Chiltern Hills. *Biulteyn Peryglacjalny*, 21, 51–73.

French, H. M. 1973: Cryopediments on the chalk of southern England. *Biulteyn Peryglacjalny*, 22, 149–56.

French, H. M. 1976: *The Periglacial Environment*. London: Longman, 309 pp.

Frey, A. 1975: River patterns in the Bristol district. In R. Peel, M. Chisholm and P. Haggett (eds), *Processes in Physical and Human Geography*, London: Heinemann, 148–65.

Frost, D. V. and Holliday, D. W. 1980: *Geology of the Country round Bellingham*. London: IGS/HMSO.

Fuller, M. A. and Reed, A. H. 1986: Rainfall, runoff and erosion on bare arable soils in East Shropshire, England. *Earth Surface Processes and Landforms* 11, 413–25.

Gale, S. J. 1981: The geomorphology of the Morecambe Bay Karst and its implications for landscape chronology. *Zeitschrift für Geomorphologie*, 25, 457–69.

Gascoyne, M., Ford, D. C. and Schwarcz, H. P. 1983: Rates of cave and land form development in the Yorkshire Dales from speleothem age dates. *Earth Surface Processes and Landforms*, 8, 557–68.

Gemmel, C., Smart, D. and Sugden, D. 1986: Striae and former ice-flow directions in Snowdonia, North Wales. *Geographical Journal*, 152, 19–29.

George, T. N. 1970: *South Wales*. London: HMSO, 152 pp.

George, T. N. 1974: Prologue to a geomorphology of Britain. *Institute of British Geographers Special Publication*, 7, 113–25.

Gerrard, A. J. 1974: 'The geomorphological importance of jointing in the Dartmoor granite. In E. H. Brown and R. S. Waters (ed.) *Progress in Geomorphology*, Institute of British Geographers Special Publications, 7, 39–51.

Gerrard, A. J. 1978: Tors and granite landforms of Dartmoor and Eastern Bodmin Moor. *Proceedings of the Ussher Society*, 4, 204–10.

Gerrard, A. J. 1988: Periglacial modification of the Cox Tor–Staple Tors area of western Dartmoor, England. *Physical Geography*, 9, 280–300.

Geyl, W. F. 1976: Tidal palaeomorphs in England. *Transactions, Institute of British Geographers*, NS 1, 203–24.

Gibbard, P. L. 1979: Middle Pleistocene drainage in the Thames Valley. *Geological Magazine*, 116, 35–44.

Gibbard, P. L. 1985: *The Pleistocene History of the Middle Thames Valley*. Cambridge: Cambridge University Press.

Gibbard, P. L., Bryant, I. D. and Hall, A. R. 1986: A Hoxnian interglacial doline infilling at Slade Oak Lane, Denham, Buckinghamshire. *Geological Magazine*, 123, 27–43.

Gibbard, P. L., Wintle, A. G. and Catt, J. A. 1987: Age and origin of clayey silt 'brickearth' in west London, England. *Journal of Quaternary Science*, 2, 3–9.

Giles R. T. and Pilkey, O. H. 1965: Atlantic beach and dune sediments of the southern United States. *Journal of Sedimentary Petrology*, 35, 900–10.

Gilman, K. and Newson, M. 1980: *Soil Pipes and Pipeflow – a Hydrological Study in Upland Wales*. Norwich: Geo Abstracts, 110 pp.

Godwin, H. 1978: *Fenland: its Ancient Past and Uncertain Future*. Cambridge: Cambridge University Press, 196 pp.

Godwin-Austen, R. A. C. 1851: Superficial accumulations on the coasts of the English Channel; and the changes they indicate. *Quarterly Journal Geological Society of London*, 7, 118–36.

Goldie, H. S. 1973: Limestone pavements of Craven. *Transactions, Cave Research Group of Great Britain*, 15, 175–89.

Goldie, H.S. 1981: Morphometry of the limestone pavements of Farleton Knott, (Cumbria, England). *Transactions British Cave Research Association*, 8, 207–24.

Goldie, H. S. 1986: Human influence on landforms: the case of limestone pavements. In K. Paterson and M. M. Sweeting (eds) *New Directions in Karst*, Norwich: Geo Books.

Gore, A. J. P. 1968: The supply of six elements by rain to an upland peat area. *Journal of Ecology*, 56, 483–95.

Gorham, E. 1958: Soluble salts in dune sands from Blakeney Point in Norfolk. *Journal of Ecology*, 46, 373–9.

Gossling, F. 1935: The structure of Bower Hill, Nutfield (Surrey). *Proceedings Geologists' Association of London*, 46, 360–90.

Gossling, F. and Bull, A. J. 1948: The structure of Tilburstow Hill, Surrey. *Proceedings Geologists' Association of London*, 59, 131–40.

Goudie, A. S. 1967: Solutional rates, processes and forms in the country between Cheltenham and Stow, Glos, unpublished BA dissertation, University of Cambridge.

Goudie, A. S. 1983: *Environmental Change*, 2nd edn. Oxford: Oxford University Press.

Goudie, A. S., Day, M. J. and Hart, M. G. 1980: Developments in the geomorphology of the Oxford Region. In T. Rowley (ed.), *The Oxford Region*, Oxford University Department for External Studies, 1–22.

Goudie, A. S. and Hart, M. 1975: Pleistocene events and forms in the Oxford Region. In C. G. Smith and D. I. Scargill (eds) *Oxford and its Region*, Oxford: Clarendon Press, 3–13.

Goudie, A. S. and Piggott, N. R. 1981: Quartzite tors, stone stripes and slopes at the Stiperstones, Shropshire, England. *Biulteyn Peryglacjalny*, 28, 47–56.

Grainger, P. and Harris, J. 1986: Weathering and slope stability on Upper Carboniferous mudrocks in south-west England. *Quarterly Journal of Engineering Geology*, 19, 155–73.

Grainger, P. and Kalaugher, P. G. 1987: Intermittent surging movements of a coastal landslide. *Earth Surface Processes and Landforms*, 12, 597–603.

Gray, A. J. 1972: The ecology of Morecambe Bay. 5. The saltmarshes of Morecambe Bay. *Journal of Applied Ecology*, 9, 207–20.

Gray, A. J. and Pearson, J. M. 1984: Spartina marshes in Poole Harbour, Dorset, with particular reference to Holes Bay. *Nature Conservancy Council, Focus on Nature Conservation*, 5, 11–14.

Gray, J. M. 1982: The last glaciers (Loch Lomond Advance) in Snowdonia, N. Wales. *Geological Journal*, 17, 111–33.

Gray, J. M. 1988a: Glaciofluvial channels below the Blakeney Esker, Norfolk. *Quaternary Newsletter*, 55, 9–12.

Gray, J. M. 1988b: Coastal cliff retreat at the Naze, Essex, since 1874: patterns, rates and processes. *Proceedings Geologists' Association*, 99, 335–8.

Green, C. P. 1973: Pleistocene river gravels and the Stonehenge problem. *Nature*, 243, 214–16.

Green, C. P. 1974: The summit surface of the Wessex Chalk. *Institute of British Geographers Special Publication*, 7, 127–38.

Green, C. P. 1985: Pre-Quaternary weathering residues, sediments and landform development: examples from southern Britain. In K. S. Richards, R. R. Arnett and S. Ellis (eds), *Geomorphology and Soils*, London: Allen & Unwin, 58–77.

Green, J. F. N. 1941: The high platforms of East Devon. *Proceedings Geologists' Association*, 52, 36–52.

Green, R. D. 1968: *Soils of Romsey Marsh*. Bulletin of Soil Survey of England and Wales, 4.

Greenwood, B. 1972: Modern analogues and the evaluation of a Pleistocene sedimentary sequence. *Transactions, Institute of British Geographers*, 56, 145–69.

Greenwood, B. 1978: Spatial variability over a beach-dune complex, North Devon, England. *Sedimentary Geology*, 21, 27–44.

Gregory, J. W. 1927: The relations of the Thames and Rhine, and age of the Strait of Dover. *Geographical Journal*, 70, 52–9.

Gregory, K. J. 1966a: Dry valleys and the composition of the drainage net. *Journal of Hydrology*, 4, 327–40.

Gregory, K. J. 1966b: Aspect and landforms in north east Yorkshire. *Biulteyn Peryglacjalny*, 15, 115–20.

Gregory, K. J. 1971: Drainage-density changes in south-west England. In K. J. Gregory and W. L. D. Ravenhill (eds), *Exeter Essays in Geography*, Exeter: University of Exeter, 33–53.

Gregory, K. J. 1976: Drainage networks and climate. In E. Derbyshire (ed.), *Geomorphology and Climate*, Chichester: John Wiley, 289–315.

Gregory, K. J. and Brown, E. H. 1966: Data processing and the study of land form. *Zeitschrift für Geomorphologie*, 10, 237–63.

Gregory, K. J. and Park, C. C. 1976: Stream channel morphology in north-west Yorkshire. *Revue de Géomorphologie Dynamique*, 25, 63–72.

Gregory, K. J. and Walling, D. E. 1968: The variation of drainage density within a catchment. *Bulletin International Association of Scientific Hydrology*, 13, 61–8.

Gresswell, R. K. 1937: The geomorphology of the south-west Lancashire coastline. *Geographical Journal*, 90, 335–49.

Gresswell, R. K. 1964: The origin of the Mersey and Dee estuaries. *Geographical Journal*, 4, 77–86.

Groom, G. E. and Williams, V. H. 1965: The solution of limestone in South Wales. *Geographical Journal*, 131, 37–41.

Guilcher, A. 1950: Nivation, cryoplanation et solifluction quaternaires dans les collines de Bretagne occidentale et du Nord de Devonshire. *Revue de Géomorphologie Dynamique*, 1, 53–78.

Gullick, C. F. W. R. 1936: A Physiographical survey of west Cornwall. *Transactions Royal Geological Society of Cornwall*, 16, 380–99.

Gunn, J. 1985: Pennine Karst areas and their Quaternary history. In R. H. Johnson (ed.), *The Geomorphology of north-west England*, Manchester: University Press, 263–81.

Hall, A. M. 1986: Deep weathering patterns in north-east Scotland and their geomorphological significance. *Zeitschrift für Geomorphologie*, 30, 407–22.

Hamblin, R. J. O. 1986: The Pleistocene sequence of the Telford District. *Proceedings Geologists' Association*, 97, 365–77.

Hancock, J. M. 1986: Cretaceous. In K. W. Glennie (ed.), *Introduction to the Petroleum Geology of the North Sea*, 2nd edn, Oxford: Blackwell Scientific Publications, 161–78.

Handa, S. and Moore, P. D. 1976: Studies in the vegetational history of mid Wales IV. Pollen analyses of some pingo basins. *New Phytologist*, 77, 203–23.

Hanwell, J. D. and Newson, M. D. 1970: The great storms and floods of July 1968 on Mendip. *Wessex Cave Club Occasional Publication*, 1.

Hardy, J. R. 1964: The movement of beach material and wave action near Blakeney Point, Norfolk. *Transactions, Institute of British Geographers*, 34, 53–69.

Hare, F. K. 1947: The geomorphology of a part of the Middle Thames. *Proceedings Geologists' Association*, 58, 29–339.

Harmsworth, G. C. and Long, S. P. 1986: An assessment of salt marsh erosion in Essex, England, with reference to the Dengie peninsula. *Biological Conservation*, 35, 377–88.

Harrod, T. M., Catt, J. A. and Weir, A. H. 1973: Loess in Devon. *Proceedings Ussher Society*, 2, 554–64.

Hartnall, T. J. 1984: Salt marsh vegetation and micro-relief development on the New Marsh at Gibraltar Point, Lincolnshire. In M. W. Clarke (ed.), *Coastal Research: UK Perspectives*, Norwich: Geobooks, 37–58.

Harvey, A. M. 1974: Gulley erosion and sediment yield in the Howgill Fells, Westmorland. In K. J. Gregory and D. E. Walling (eds), *Fluvial Processes in Instrumented Watersheds*, Institute of British Geographers Special Publication, 6, 45–58.

Harvey, A. M., Oldfield, F., Baron, A. F. and Pearson, G. W. 1981:

Dating of post-glacial landforms in the central Howgills. *Earth Surface Processes and Landforms*, 6, 401–12.

Harvey, A. M. and Renwick, W. H. 1987: Holocene alluvial fan and terrace formation in the Bowland Fells, Northwest England. *Earth Surface Processes and Landforms*, 12, 249–57.

Harwood, D. 1988: Was there a glacial Lake Harrison in the south Midlands of England? *Mercian Geologist*, 11 (3), 145–53.

Hawkins, A. B. and Kellaway, G. A. 1971: Field meeting at Bristol and Bath with special reference to new evidence of glaciation. *Proceedings Geologists' Association*, 82, 267–92.

Hawkins, A. B. and Privett, K. D. 1979: Engineering geomorphological mapping as a technique to elucidate areas of superficial structures; with examples from the Bath area of the South Cotswolds. *Quarterly Journal of Engineering Geology*, 12, 221–33.

Hay, J. T. C. 1978: Structural development in the Northern North Sea. *Journal of Petroleum Geology*, 1, 65–77.

Hay, Th. 1936: Stone stripes. *Geographical Journal*, 87, 47–50.

Hay, Th. 1937: Physiographical notes on the Ullswater area. *Geographical Journal*, 90, 426–45.

Hay, T. 1943: Notes on glacial erosion and stone stripes. *Geographical Journal*, 13, 13–20.

Hepburn, I. 1944: The vegetation of the sand dunes of the Camel estuary. *Journal of Ecology*, 32, 180–92.

Hey, R. W. 1958: High level gravels in and near the lower Severn valley. *Geological Magazine*, 95, 161–8.

Higgins, L. C. 1933: An investigation into the problem of the sand dune areas on the South Wales coast. *Archaeologia Cambrensis*, 88, 26–67.

Hills, R. C. 1971: The influence of land management and soil characteristics on infiltration and the occurrence of overland flow. *Journal of Hydrology*, 13, 163–81.

Hindley, R. 1965: Sink-holes on the Lincolnshire limestone between Grantham and Stamford. *East Midland Geographer*, 3, 454–60.

Hodgson, E. 1867: The moulded limestones of Furness. *Geological Magazine*, 4, 401–6.

Hodgson, J. M., Rayner, J. H. and Catt, J. A. 1974: The geomorphological significance of the Clay-with-flints on the South Downs. *Transactions, Institute of British Geographers*, 61, 119–29.

Hollingworth, S. E. 1929: The evolution of the Eden drainage in the south and west. *Proceedings Geologists' Association*, 40, 115–38.

Hollingworth, S. E. 1934: Some solifluction phenomena in the northern part of the Lake District. *Proceedings Geologists' Association*, 45, 167–88.

Hollingworth, S. E. 1935: High level erosional platforms in Cumberland. *Proceedings Yorkshire Geological Society*, 23, 159–77.

Hollingworth, S. E. 1938: The recognition and correlation of high level erosion surfaces in Britain: a statistical study. *Quarterly Journal of the Geological Society of London*, 94, 55–84.

Hollingworth, S. E., Taylor, J. H. and Kellaway, G. A. 1944: Large-scale superficial structures in the Northampton Ironstone Field. *Quarterly Journal of the Geological Society of London*, 100, 1–44.

Hollis, G. E. and Luckett, J. K. 1976: The response of natural river channels to urbanisation: two case studies from south-east England. *Journal of Hydrology*, 30, 351–63.

Holmes, S. C. A. 1981: *Geology of the Country around Faversham*. London. IGS/HMSO.

Hooke, J. M. 1980: Magnitude and distribution of rates of river bank erosion. *Earth Surface Processes*, 5, 143–57.

Horswill, P. and Horton, A. 1976: Cambering and valley bulging in the Gwash valley at Empingham, Rutland. *Philosophical Transactions of the Royal Society of London* A, 283, 427–62.

Horton, A., Worssam, B. C. and Whittow, J. B. 1981: The Wallingford Fan Gravel. *Philosophical Transactions Royal Society of London*, 293 B, 215–55.

Horton, R. E. 1945: Erosional development of streams and their drainage basins: hydrophysical approach to quantitative morphology. *Bulletin, Geological Society of America*, 56, 275–370.

Howe, G. M., Slaymaker, H. O. and Harding, D. M. 1966: Flood hazard in Mid-Wales. *Nature*, 212, 584–5.

Howell, F. T. and Jenkins, P. L. 1976: Some aspects of the subsidences in the rock salt districts of Cheshire, England. *Publication of the International Association of Hydrological Sciences*, 121, 507–20.

Hubbard, J. C. E. and Stebbings, R. E. 1968: *Spartina* marshes in Southern England. VII Stratigraphy of the Keysworth marsh, Poole Harbour. *Journal of Geology*, 56, 707–22.

Hutchinson, J. N. 1968: Field meeting on the coastal landslides of Kent. *Proceedings Geologists' Association*, 79, 227–37.

Hutchinson, J. N. 1971: Field and laboratory studies of a fall in Upper Chalk cliffs at Jess Bay, Isle of Thanet. Paper presented at Roscoe Memorial symposium, Cambridge, 12 pp.

Hutchinson, J. N. 1980a: Possible late Quaternary pingo remnants in central London. *Nature*, 284, 253–5.

Hutchinson, J. N. 1980b: The record of peat wastage in the East Anglian Fenland at Holme Post, AD 1948–1978. *Journal of Ecology*, 68, 229–49.

Hutchinson, J. N. 1983: A pattern in the incidence of major coastal landslides. *Earth Surface Processes and Landforms*, 8, 391–7.

Hutchinson, J. N. 1987: Some coastal landslides of the southern Isle of Wight. In K. E. Barber (ed.), *Wessex and the Isle of Wight Field Guide*, Cambridge: Q.R.A., 123–35.

Hutchinson, J. N., Bromhead, E. N. and Lupini, J. F. 1980: Additional observations on the Folkestone Warren landslides. *Quarterly Journal of Engineering Geology*, 13, 1–31.

Hutchinson, J. N. and Gostelow, T. P. 1976: The development of an abandoned cliff in London Clay at Hadleigh, Essex. *Philosophical Transactions Royal Society of London* A, 283, 557–604.

Hutchinson, J. N., Somerville, S. H. and Petley, D. J. 1973: A landslide in periglacially disturbed Etruria Marl at Bury Hill, Staffordshire. *Quarterly Journal of Engineering Geology*, 6, 377–404.

Imeson, A. C. 1974: The origin of sediment in a moorland catchment with particular reference to the role of vegetation. *Institute of British Geographers' Special Publication*, 6, 59–72.

Innes, J. L. 1983: Lichenometric dating of debris-flow deposits in the Scottish Highlands. *Earth Surface Processes and Landforms*, 8, 579–88.

Institute of Hydrology: Research Report, 1984–1987.

Isaac, K. P. 1981: Tertiary weathering profiles in the plateau deposits of East Devon. *Proceedings Geologists' Association*, 92, 159–68.

Isaac, K. P. 1983: Tertiary lateritic weathering in Devon, England, and the Palaeogene continental environment of South-west England. *Proceedings Geologists' Association*, 94, 105–14.

James, A. N., Cooper, A. H. and Holliday, D. W. 1981: Solution of the gypsum cliff (Permian, Middle Marl) by the River Ure at Ripon Parks, North Yorkshire. *Proceedings Yorkshire Geological Society*, 43, 433–50.

Jarzembowski, E. A. 1980: Fossil insects from the Bembridge Marls, Palaeogene of the Isle of Wight, Southern England. *Bulletin of the British Museum (Natural History)*, Geology Series, 33(4), 237–93.

Jaynes, S. M. and Cooke, R. U. 1987: Stone weathering in southeast England. *Atmospheric Environment*, 21 (7), 1601–22.

Jenkins, C. A. and Vincent, A. 1981: Periglacial features in the Bovey Basin. *Proceedings Ussher Society*, 5, 200–5.

Jennings, J. N. 1952: The origin of the Broads. *Royal Geographical Research Series*, No. 2.

Jennings, J. N. 1985: *Karst Geomorphology*. Oxford: Basil Blackwell.

Job, D. A. 1982: Runoff and sediment output from a small lowland catchment – the example of Preston Montford Brook, Shropshire. *Field Studies*, 5, 685–729.

Johnson, R. H. 1975: Some late Pleistocene involutions at Dalton-in-Furness, northern England. *Geological Journal*, 10, 23–34.

Johnson, R. H. 1980: Hillslope stability and landslide hazard – a case study from Longdendale, north Derbyshire, England. *Proceedings Geologists' Association*, 91 (4) 315–25.

Johnson, R. H. 1985: The imprint of glaciation on the west Pennine Uplands. In R. H. Johnson (ed.), *The Geomorphology of North-west England*, Manchester: Manchester University Press, 237–62.

Johnson, R. H. 1987: Dating of ancient, deep-seated landslides in temperate regions. In M. G. Anderson and K. S. Richards (eds), *Slope Stability*, Chichester: John Wiley, 561–600.

Johnson, R. H. and Rice, R. J. 1961: Denudation chronology of the Southwest Pennines. *Proceedings Geologists' Association*, 72, 21–32.

Johnson, R. H. and Vaughan, R. D. 1983: The Alport Castles, Derbyshire:

a south Pennine slope and its geomorphic history. *East Midland Geographer*, 8, 79–88.

Jones, D. K. C. 1974: The influence of the Calabrian transgression on the drainage evolution of south-east England. *Institute of British Geographers Special Publication*, 7, 139–58.

Jones, D. K. C. (ed.) 1980: *The Shaping of Southern England*. London: Academic Press.

Jones, D. K. C. 1981: *Southeast and Southern England*. London: Methuen.

Jones, D. K. C. 1983: Human occupance and the physical environment. In R. J. Johnston and J. C. Doornkamp (eds), *The Changing Geography of the United Kingdon*, London: Methuen, 327–61.

Jones, D. K. C. 1985: Shaping the land: The geomorphological background. In S. R. J. Woodell (ed.), *The English Landscape Past, Present, and Future*, Oxford: Oxford University Press, 4–47.

Jones, D. K. C., Griffiths, J. S. and Lee, E. M. 1988: The distribution of recorded landslides in south west England. *Proceedings of the Ussher Society*, 7, 91–92.

Jones, J. A. A. 1968: Morphology of the Lapworth Valley, Warwickshire. *Geographical Journal*, 134 (2), 215–26.

Jones, J. A. A. 1981: *The Nature of Soil Piping – a Review of Research*. Norwich: Geobooks, 301 pp.

Jones, M. E., Allison, R. J. and Gilligan, J. 1983: On the relationships between geology and coastal landforms in central Southern England. *Proceedings of the Dorset Natural History and Archaeological Society*, 105, 107–18.

Jones, O. T. 1931: Some episodes in the geological history of the Bristol Channel region. *Report of the British Association for the Advancement of Science*, 57–82.

Jones, P. F. 1979: The origin and significance of dry valleys in South East Derbyshire. *Mercian Geologist*, 7, 1–18.

Jones, P. F. and Weaver, J. D. 1975: Superficial valley folds of late Pleistocene age in the Breadsall area of South Derbyshire. *Mercian Geologist*, 5, 279–290.

Jones, R., Benson-Evans, K. and Chambers, F. M. 1985: Human influence upon sedimentation in Llangorse Lake, Wales. *Earth Surface Processes and Landforms*, 10(3), 227–35.

Jones, R. J. 1965: Aspects of the biological weathering of limestone pavement. *Proceedings Geologists' Association*, 76, 421–33.

Jukes-Browne, A. J. 1906: The clay-with-flints; its origin and distribution. *Quarterly Journal of the Geological Society of London*, 62, 132–64.

Keen, D. H. 1985: Late Pleistocene Deposits and Mollusca from Portland, Dorset. *Geological Magazine*, 122, 181–6.

Kellaway, G. A. 1971: Glaciation and the stones of Stonehenge. *Nature*, 232, 30–5.

Kellaway, G. A. 1972: Development of non-diastrophic Pleistocene structures in relation to climate and physical relief in Britain. *Proceedings*

24th International Geological Congress, Section 12, 136–46.

Kellaway, G. A., Horton, A. and Poole, E. G. 1971: The development of some Pleistocene structures in the Cotswolds and Upper Thames basin. *Bulletin Geological Survey of Great Britain*, 37, 1–28.

Kellaway, G. A., Redding, J. H., Shephard-Thorn, E. R. and Destombes, J. P. 1975: The Quaternary history of the English Channel. *Philosophical Transactions Royal Society of London* A, 278, 189–218.

Kellaway, G. A. and Taylor, J. H. 1953: Early stages in the physiographic evolution of a portion of the east Midlands. *Quarterly Journal Geological Society of London*, 108, 343–75.

Kendall, P. F. 1902: A system of glacier-lakes in the Cleveland Hills. *Quarterly Journal of the Geological Society*, 58, 471–571.

Kennard, A. S. and Warren, S. H. 1903: The blown sands and associated deposits of Tower Head, near Newquay, Cornwall. *Geological Magazine*, IV, 10, 19–25.

Kent, P. 1980: *Eastern England from the Tees to the Wash*, 2nd edn. London: Institute of Geological Sciences.

Kerney, M. P., Brown, E. H. and Chandler, T. J. 1964: The late-glacial and post-glacial history of the Chalk escarpment near Brook, Kent. *Philosophical Transactions of the Royal Society of London* B, 248, 135–204.

Kidson, C. 1950: Dawlish Warren: a study of the evolution of the sand spits across the north of the River Exe in Devon. *Transactions, Institute of British Geographers*, 16, 69–80.

Kidson, C 1960: The shingle complexes of Bridgwater Bay. *Transactions, Institute of British Geographers*, 28, 78.

Kidson, C. 1962: The denudation chronology of the River Exe. *Transactions, Institute of British Geographers*, 31, 43–66.

Kidson, C. 1963: The growth of sand and shingle spits across estuaries. *Zeitschrift für Geomorphologie*, 7, 1–22.

Kidson, C. and Carr, A. P. 1960: Dune reclamation at Braunton Burrows. *Chartered Surveyor*, 93, 298–303.

Kidson, C. and Tooley, M. J. (eds) 1977: *The Quaternary History of the Irish Sea*. Liverpool: Seel House Press.

King, C. A. M. 1960: *The Yorkshire Dales*. Sheffield: Geographical Association, 29 pp.

King, C. A. M. 1963: Some problems concerning marine planation and the formation of erosion surfaces. *Transactions, Institute of British Geographers*, 33, 29–43.

King, C. A. M. 1976: *Northern England*. London: Methuen, 213 pp.

King, C. A. M. 1978: Changes in the foreshore and spit between 1972 and 1978 at Gibraltar Point, Lincolnshire. *East Midland Geographer*, 7, 73–82.

King, C. A. M., McCullagh, M. J. 1971: A simulation model of a complex recurved spit. *Journal of Geology*, 79, 22–37.

King, L. C. 1953: Canons of landscape evolution. *Bulletin Geological Society of America*, 64, 721–52.

Kirkaldy, J. F. 1950: Solution of the chalk in the Mimms Valley, Herts. *Proceedings Geologists' Association*, 61, 219–23.

Kirkby, M. J. 1984: Modelling cliff development in South Wales: Savigear re-viewed. *Zeitschrift für Geomorphologie*, 28, 405–26.

Kirkby, M. J. and Chorley, R. J. 1967: Throughflow, overland flow and erosion. *Bulletin International Association of Scientific Hydrology*, 12, 5–21.

Knowles, A. 1985: The Quaternary history of North Staffordshire. In Johnson, R. H. (ed.), *The Geomorphology of North-West England*, Manchester: Manchester University Press, 222–36.

Ladle, M. (ed.) 1981: *The Fleet and Chesil Beach*. Dorset County Council. 74 pp

Lambert, J. H., Jennings, J. N., Smith, C. T., Green, C. and Hutchinson J. N. 1970: The making of the Broads: a reconsideration of their origin in the light of new evidence. *Royal Geographical Society Research Series*, 3.

Lambert, J. M. and Davies, M. R. 1940: A sandy area in the Dovey estuary. *Journal of Ecology*, 28, 453.

Lapworth, C. and Watts, W. W. 1910: Shropshire. In H. W. Monckton and R. Herries (eds), *Geology in the Field*, London: Stanford, 739–69.

Lawler, D. M. 1986: River bank erosion and the influence of frost: a statistical examination. *Transactions, Institute of British Geographers*, 11, 227–42.

Lawler, D. M. 1987: Climatic change over the last millenium in Central Britain. In K. J. Gregory, J. Lewin and J. B. Thornes (eds), *Palaeohydrology in Practice*, 99–129. John Wiley, Chichester.

Leach, A. L. 1933: The geology and scenery of Tenby and the south Pembrokeshire coast. *Proceedings Geologists' Association*, 44, 187–216.

Lee, M. P. 1979: Loess from the Pleistocene of the Wirral Peninsula. *Proceedings Geologists' Association*, 50, 21–6.

Lees, D. J. 1982: The sand dunes of Gower as potential indicators of climatic change in historical time. *Cambria*, 9, 23–5.

Lewin, J. 1966: Fossil ice wedges in Hampshire. *Nature*, 211, 728.

Lewin, J. 1969: The Yorkshire wolds: a study in geomorphology. *University of Hull Occasional Paper*, 11, 89 pp.

Lewin, J., Brasley, S. B. and Mackin, M. G. 1983: Historical valley alluviation in mid-Wales. *Geological Journal*, 18, 331.

Lewin, J. R., Cryer, R. and Harrison, D. I. 1974: Sources of sediments and solutes in mid-Wales. *Institute of British Geographers Special Publication*, 6, 73–85.

Lewis, C. A. 1970: In C. A. Lewis (ed.), *The Glaciations of Wales and Adjoining Regions*, London: Longman, 147–73.

Lewis, K. 1983: A morphometric and geological study of limestone pavements in South Wales. *Cave Science*, 10, 199–204.

Ley, R. G. 1979: The development of marine karren along the Bristol Channel coastline. *Zeitschrift für Geomorphologie Supplementband*, 32, 75–89.

Liddle, M. J. 1975: A selective review of the ecological effects of human trampling on natural ecosystems. *Biological Conservation*, 7, 17–36.

Liddle, M. J. and Greig-Smith, P. 1975: A survey of tracks and paths in a sand dune ecosystem. *Journal of Applied Ecology*, 12, 893–930.

Linton, D. L. 1951: Midland drainage: some considerations bearing on its origin. *Advancement of Science*, 7 (28), 449–56.

Linton, D. L. 1955: The problem of tors. *Geographical Journal*, 121, 470–87.

Linton, D. L. 1957: Radiating valleys in glaciated lands. *Tijdschrift van het Koninklijke Nederland Aardrijkskundig Genootschap*, 74, 297–312.

Linton, D. L. 1963: The forms of glacial erosion. *Transactions, Institute of British Geographers*, 33, 1–18.

Livingstone, I. 1980: Pits and ponds in Kent. *Bygone Kent*, 1, 161–65.

Lockwood, J. G. 1982: Snow and ice balance in Britain at the present time and during the last glacial maximum and late glacial periods. *Journal of Climatology*, 2, 209–31.

Long, D. and Stoker, M. S. 1986: Valley asymmetry: evidence for periglacial activity in the Central North Sea. *Earth Surface Processes and Landforms*, 11, 525–32.

Lovell, J. P. B. 1986: Cenozoic. In K. W. Glennie (ed.), *Introduction to the Petroleum Geology of the North Sea*, 2nd edn, Oxford: Blackwell Scientific Publications, 179–96.

Lowe, J. J. and Walker, M. J. C. 1984: *Reconstructing Quaternary Environments*. London: Longman.

Lyell, C. 1835: *Principles of Geology*, Vol. III. London: Murray, 455 pp.

McArthur, J. L. 1977: Quaternary erosion in the Upper Derwent basin and its bearing on the age of surface features in the Southern Pennines. *Transactions, Institute of British Geographers*, 2, 490–7.

McArthur, J. L. 1981: Periglacial slope planations in the Southern Pennines. *Biulteyn Peryglacjalny*, 28, 85–97.

McConnell, R. B. 1939a: The relic surfaces of the Howgill Fells. *Proceedings of the Yorkshire Geological Society*, 24, 152–64.

McConnell, R. B. 1939b: Residual erosion surfaces in mountain ranges. *Proceedings of the Yorkshire Geological Society*, 24, 76–98.

Mackinder, H. J. 1904: *Britain and the British Seas*. London: Heinemann.

Mackinder, H. J. 1930: *Britain and the British Seas*, 2nd edn. Oxford: Clarendon Press.

Mackintosh, D. 1869: *The Scenery of England and Wales, its Character and Origin*. London: Longmans and Green Co., 399 pp.

Macklin, M. G. and Lewin, J. 1986: Terraced Hills of Pleistocene and Holocene age in the Rheidol Valley, Wales. *Journal of Quaternary Science*, 1, 21–34.

Marker, M. E. 1967: The Dee Estuary: its progressive silting and salt marsh development. *Transactions, Institute of British Geographers*, 41, 65–71.

Marr, J. E. 1916: *The Geology of the Lake District*. Cambridge: Cambridge University Press, 220 pp.

Marr, J. E. and Fearnsides, W. G. 1909: The Howgill Fells and their topography. *Journal of the Geological Society of London*, 65, 587–610.

Marshall, J. R. 1962: The morphology of the Upper Solway salt marshes. *Scottish Geographical Magazine*, 78, 81–99.

Matthews, B. 1970: Age and origin of aeolian sand in The Vale of York. *Nature*, 227, 1234–36.

May, V. J. 1966: A preliminary study of recent coastal change and sea defences in South East England. *Southampton Research Series in Geography*, 3, 3–24.

May, V. and Heeps, C. 1985: The nature and rates of change on chalk coastlines. *Zeitschrift für Geomorphologie Supplement band*, 57, 81–94.

Merefield, J. R. 1984: Modern cool-water beach sands of south-west England. *Journal of Sedimentary Petrology*, 54, 413–24.

Miller, A. A. 1937: The 600-foot plateau in Pembrokeshire and Carmarthenshire. *Geographical Journal*, 90, 148–59.

Miller, A. A. 1938: Pre-glacial erosion surfaces round the Irish Sea. *Proceedings Yorkshire Geological Society*, 24, 31–59.

Mitchell, D. J. and Gerrard, A. J. 1987: Morphological response and sediment patterns. In K. J. Gregory, J. Lewin and J. B. Thornes (eds) *Palaeohydrology in Practice*. Chichester: John Wiley, 177–99.

Mitchell, G. F. 1977: Raised beaches and sea-levels. In F. W. Shotton (ed.), *British Quaternary Studies*, Oxford: Clarendon Press, 169–86.

Mitchell, G. F. and Orme, A. R. 1967: The Pleistocene deposits of the Isles of Scilly. *Quarterly Journal, Geological Society of London*, 123, 59–92.

Mitchell, G. F., Penny, L. F., Shotton, F. S. and West R. G. 1973: *A correlation of Quaternary Deposits in the British Isles*. London: Geological Society.

Moffatt, A. J., Catt, J. A., Webster, R. and Brown, E. H. 1986: A re-examination of the evidence for a Plio-Pleistocene marine transgression on the Chiltern Hills. 1. Structures and Surfaces. *Earth Surface Processes and Landforms*, 11 (1), 95–106.

Moffatt, A. J. and Catt, J. A. 1986a: A re-examination of the evidence for a Plio-Pleistocene marine transgression on the Chiltern Hills. II. Drainage Patterns. *Earth Surface Processes and Landforms*, 11, 169–80.

Moffatt, A. J. and Catt, J. A. 1986b: A re-examination of the evidence for a Plio-Pleistocene marine transgression on the Chiltern Hills. III. Deposits. *Earth Surface Processes and Landforms*, 11, 233–47.

Moon, H. P. and Green, F. H. W. 1940: Water meadows in southern England. *Annex to Part 89, Report of the Land Utilisation Survey of Britain*, 373–90.

Moore, E. J. 1931: The ecology of the Ayreland of Bride, Isle of Man. *Journal of Ecology*, 19, 115–36.

Moore, R. 1986: The Fairlight landslips: The location, form and behaviour of coastal landslides with respect to toe erosion. *Occasional Paper*, 27, Dept. of Geography, King's College, London, 43 pp.

Moore, R. J. and Newson, M. D. 1986: Production, storage and output of coarse upland sediments: natural and artificial influences as revealed by research catchment studies. *Journal of the Geological Society* (London), 143(6), 921–6.

Morgan, A. V. 1971: Polygonal patterned ground of late Weichselian age in the area North and West of Wolverhampton, England. *Geografiska Annaler*, 53A, 146–56.

Morgan, R. P. C. 1971: A morphometric study of some valley systems in the English chalklands. *Transactions, Institute of British Geographers*, 54, 33–44.

Morgan, R. P. C. 1977: Soil erosion in the United Kingdom: field studies in the Silsoe Area, 1973–75. *Occasional Paper*, No. 4, National College of Agricultural Engineering, 41 pp.

Morgan, R. P. C. 1986: *Soil Erosion and Conservation*. London: Longman.

Morris, L. 1974: The geology and geomorphology of the Malvern Hills and their surroundings. In B. H. Adlam (ed.), *Worcester and its Region*, Worcester: Geographical Association, 87–96.

Moseley, F. 1961: Erosion surfaces in the Forest of Bowland. *Proceedings Yorkshire Geological Society*, 33, 173–96.

Mosley, M. P. 1975: Channel changes in the River Bollin, Cheshire, 1872–1973. *East Midland Geographer*, 6, 185–99.

Moss, B. 1984: Mediaeval man-made lakes: progeny and casualties of English social history, parents of twentieth century ecology. *Transactions Royal Society of South Africa*, 45 (2), 115–28.

Mottershead, D. N. 1967: The evolution of Valley of Rocks and its landforms. *Exmoor Review*, 8, 69–72.

Mottershead, D. N. 1971: Coastal head deposits between Start Point and Hope Cave, Devon. *Field Studies*, 3, 433–53.

Mottershead, D. N. 1982: Coastal spray weathering of bedrock in the supratidal zone of the east Prawle, South Devon. *Field Studies*, 5, 663–84.

Mottershead, D. N., Gilbertson, D. D. and Keen, D. H. 1987. The raised beaches and shore platforms of Tor Bay: a re-evaluation. *Proceedings Geologists' Association*, 98, 241–57.

Murchison, R. I. 1854: *Siluria*. London: Murray, 523 pp.

Murgatroyd, A. L. and Ternan, J. L. 1983: The impact of afforestation on stream bank erosion and channel form. *Earth Surface Processes and Landforms*, 8, 357–69.

Naylor, D. and Shannon, P. 1982: *Geology of Offshore Ireland and West Britain*. London: Graham & Trotman, 161 pp.

Neal, C., Smith, C. J., Walls, J. and Dunn, C. S. 1986: Major, minor and trace element mobility in the acidic upland forested catchment of the upper River Severn, Mid Wales. *Journal of the Geological Society (London)*, 143, 635–48.

Neilson, G., Musson, R. M. W. and Burton, P. W. 1984: The 'London' Earthquake of 1580, April 6. *Engineering Geology*, 20, 113–41.

Newson, M. D. 1971: A model of subterranean limestone erosion in the British Isles. *Transactions, Institute of British Geographers*, 54, 55–70.

Newson, M. D. 1975: *Flooding and Flood Hazard in the United Kingdom*. Oxford: Oxford University Press.

Newson, M. D. 1978: Drainage basin characteristics, their selection, derivation and analysis for a flood study of the British Isles. *Earth Surface Processes*, 3, 277–93.

Newson, M. 1980: The geomorphological effectiveness of floods – a contribution stimulated by two recent events in mid-Wales. *Earth Surface Processes*, 5, 1–16.

Newson, M. D. 1981: Mountain streams. In Lewin, J. (ed.), *British Rivers*, London: Allen & Unwin, 59–89.

Newson, M. D. 1986: River basin engineering – fluvial geomorphology. *Journal of the Institution of Water Engineers and Scientists*, 40, 309–25.

O'Connor, J., Williams, D. S. F. and Davies, G. M. 1974: Karst features of Malham and the Craven Fault Zone. In A. C. Waltham (ed.), *The Limestones and Caves of North-West England*, Newton Abbott: David and Charles, 395–409.

Ollier, C. D. and Thomasson, A. J. 1957: Asymmetrical valleys of the Chiltern Hills. *Geographical Journal*, 122, 71–80.

Orford, J. 1987: Coastal processes: the coastal response to sea-level variation. In R. J. N. Devoy (ed.) *Sea Surface Studies*. London: Croom Helm, 415–63.

Orme, A. R. 1962: Abandoned and composite sea cliffs in Britain and Ireland. *Irish Geography*, 4, 279–91.

O'Sullivan, P. E., Coard, M. A. and Pickering, D. A. 1982: The use of laminated lake sediments in the estimation and calibration of erosion rates. *Publication, International Association of Hydrological Science*, 137, 385–96.

Owen, T. R. 1976: *The Geological Evolution of the British Isles*. Oxford: Pergamon. 161 pp.

Oxford, S. P. 1985: Protalus ramparts, protalus rock glaciers and soliflucted till in the northwest part of the English Lake District. In J. Boardman (ed.) *Field Guide to the Periglacial Landforms of Northern England*. Cambridge: Quaternary Research Association, 38–46.

Oxley, N. C. 1974: Suspended sediment delivery rates and solute concentration of stream discharge in two Welsh catchments. *Institute of British Geographers' Special Publication*, 6, 141–53.

Packham, J. R. and Liddle, M. J. 1970: The Cefni salt marsh, Anglesey and its recent development. *Field Studies*, 3, 331–56.

Page, H. 1982: Some notes on the geomorphological and vegetational history of the Saltings at Brean. *Somerset Archaeology and Natural History*, 1982, 120–25.

Palmer, J. 1956: Tor formation at the Bridestones in north-east Yorkshire and its significance in relation to problems of valley-side development and regional glaciation. *Transactions, Institute of British Geographers*, 22, 55–71.

Palmer, J. and Neilson, R. A. 1962: The origin of granite tors on Dartmoor, Devonshire. *Proceedings Yorkshire Geological Society*, 33, 315–40.

Park, C. C. 1979: Tin streaming and channel changes: some, preliminary observations from Dartmoor, England. *Catena*, 6, 235–44.

Parks, D. A. and Rendell, H. M. 1988: TL dating of brickearths from SE England. *Quaternary Science Reviews*, 7, 305–8.

Parrish, J. T. 1987: Global palaeogeography and palaeoclimate of the Late Cretaceous and Early Tertiary. In E. M. Friis, W. G. Chaloner and P. R. Crane (eds), *The Origins of Angiosperms and their Biological Consequences*. Cambridge: Cambridge University Press, 358 pp.

Parry, J. T. 1960a: The erosion surfaces of the south western Lake District. *Transactions, Institute of British Geographers*, 28, 39–54.

Parry, J. T. 1960b: Limestone pavements of north western England. *Canadian Geographer*, 16, 14–21.

Paterson, K. 1970: Aspects of the geomorphology of the Oxford Region, unpublished PhD thesis, University of Oxford.

Paterson, K. 1977: Scarp-face dry valleys, near Wantage, Oxfordshire. *Transactions, Institute of British Geographers*, NS 2, 192–204.

Paterson, T. T. 1940: The effects of frost action and solifluction around Baffin Bay, and the Cambridge district. *Quarterly Journal, Geological Society of London*, 96, 99–130.

Pattison, J. and Williamson, I. T. 1986: The saltern mounds of north-east Lincolnshire. *Proceedings of the Yorkshire Geological Society*, 46, 77–79.

Paul, M. A. 1983: The supraglacial land system. In N. Eyles (ed.), *Glacial Geology*, Oxford: Pergamon Press, 70–90.

Pearsall, W. H. N. 1934: North Lancashire sand dunes. *Naturalist*, 1934, 201–5.

Pedley, H. M. 1987: The Flandrian (Quaternary) Caerwys Tufa, North Wales: an ancient barrage tufa deposit. *Proceedings Yorkshire Geological Society*, 46 (2), 141–52.

Peel, R. F. and Palmer, J. 1955: The physiography of the Vale of York. *Geography*, 40, 215–27.

Peltier, L. C. 1959: The geographical cycle in periglacial regions as it is related to climatic geomorphology. *Annals of the Association of American Geographers*, 40, 214–36.

Pemberton, M. 1980: Earth hummocks at low elevation in The Vale of Eden, Cumbria. *Transactions, Institute of British Geographers*, New Series 5, 487–501.

Penn, S., Royce, C. J. and Evans, C. J. 1983: The periglacial modification of the Lincoln Scarp. *Quarterly Journal of Engineering Geology, London*, 16, 309–18.

Penning-Rowsell, E. C. and Townshend, J. R. G. 1978: The influence of scale on the factors affecting stream channel slope. *Transactions, Institute of British Geographers*, NS 3, 395–415.

Pennington, W. 1981: Records of a lake's life in time: the sediments. *Hydrobiologia*, 79, 197–219.

Pentecost, A. 1978: Blue-green algae and freshwater carbonate deposits. *Proceedings of the Royal Society of London*, B 200, 43–61.

Pentecost, A. 1981: The tufa deposits of the Malham District, North Yorkshire. *Field Studies*, 5, 365–87.

Pentecost, A. and Lord, T. 1988: Postglacial tufas and travertines from the Craven District of Yorkshire. *Cave Science*, 15, 15–19.

Permadasa, M. A., Greig-Smith, P. and Lovell, P. A. 1974: A quantitative description of the distribution of annuals in the dune system at Aberffraw, Anglesey. *Journal of Ecology*, 62, 379–407.

Perrin, R. M. S. 1955: Studies in pedogenesis, unpublished PhD thesis, University of Cambridge.

Perry, A. 1981: *Environmental Hazards in the British Isles*. London: Allen & Unwin.

Pethick, J. A. 1980: Salt-marsh initiation during the Holocene transgression: the example of the North Norfolk marshes, England. *Journal of Biogeography*, 7, 1–9.

Pethick, J. S. 1981: Long term accretion rates on tidal salt marshes. *Journal of Sedimentary Petrology*, 51, 571–7.

Pethick, J. S. 1984: *An Introduction to Coastal Geomorphology*. London: Edward, Arnold.

Petts, G. E. 1979: Complex response of river channel morphology subsequent to reservoir construction. *Progress in Physical Geography*, 3, 329–62.

Pigott, C. D. 1965: The structure of limestone surfaces in Derbyshire. *Geographical Journal*, 131, 41–4.

Pissart, A. 1963: Les traces de 'pingos' du Pays de Galles (Grande-Bretagne) et du Plateau des Hautes Fagnes (Belgique). *Zeitschrift für Geomorphologie*, 7, 147–65.

Pitman, J. I. 1986: Chemical weathering of the East Yorkshire Chalk. In K. Paterson and M. M. Sweeting (eds), *New Directions in Karst*, Norwich: Geobooks, 77–113.

Pitts, J. 1983: The temporal and spatial development of landslides in the Axmouth-Lyme Regis undercliffs National Nature Reserve, Devon. *Earth Surface Processes and Landforms*, 8(6), 589–603.

Pitts, J. and Brundsden, D. 1987: A reconsideration of the Bindon landslide of 1839. *Proceedings Geologists' Association*, 98, 1–18.

Pitty, A. F. 1968: The scale and significance of solutional loss from the limestone tract of the southern Pennines. *Proceedings Geologists' Association*, 79, 153–77.

Pizzey, J. M. 1975: Assessment of dune stabilisation at Camber, Sussex, using air photographs. *Biological Conservation*, 7, 275–8.

Plant, J. 1866: On the existence of a sea beach on the limestone moors near Buxton. *Transactions, Manchester Geological Society*, 5, 272.

Plint, A. G. 1983: Sandy fluvial point-bar sediments from the Middle Eocene of Dorset, England. *Special Publications International Association of Sedimentologists*, 6, 355–68.

Pollard, E. and Miller, A. 1968: Wind erosion in the East Anglian Fens. *Weather*, 23, 414–17.

Poole, E. G. and Whiteman, A. J. 1961: The glacial drifts of the southern part of the Shropshire–Cheshire basin. *Quarterly Journal of the Geological Society*, 117, 91–130.

Potts, A. S. 1971: Fossil cryonival features in central Wales. *Geografiska Annaler*, 53A, 39–51.

Potts, E. A. 1968: The geomorphology of the sand dunes of South Wales with special reference to Gower, unpublished PhD thesis, University College of Wales, Swansea.

Prentice, J. E. and Morris, P. G. 1959: Cemented screes in the Manifold valley, North Staffordshire. *East Midlands Geographer*, 2 (11), 16–19.

Prestwich, J. 1864: On the loess of the south of England, and of the Somme and the Seine. *Philosophical Transactions Royal Society*, 154, 247–309.

Price, R. J. 1973: *Glacial and Fluvioglacial Landforms*. Edinburgh: Oliver and Boyd.

Prince, H. C. 1962: Pits and ponds in Norfolk. *Erdkunde*, 16, 10–31.

Prince, H. C. 1964: The origin of pits and depressions in Norfolk. *Geography*, 49, 15–32.

Pullan, R. A. 1959: Notes on periglacial phenomena: tors. *Scottish Geographical Magazine*, 75, 51–5.

Rackham, O. 1986: *The History of the Countryside*. London: Dent.

Radley, J. 1962: Peat erosion on the high moors of Derbyshire and west Yorkshire. *East Midlands Geographer*, 3(1), 40–50.

Radley, J. and Sims, C. 1967: Wind erosion in East Yorkshire. *Nature*, 216, 20–2.

Ramsay, A. C. 1846: The denudation of South Wales. *Memoir Geological Survey of Great Britain*, 1.

Ramsay, A. C. 1860: *The Old Glaciers of Switzerland and North Wales*. London: Longman, 116 pp.

Ramsay, A. C. 1878: *The Physical Geology and Geography of Great Britain*, 5th edn. London: Stanford, 639 pp.

Randall, H. J. 1961: *The Vale of Glamorgan – Studies in Landscape and History*. Newport: Johns.

Randall, R. E. 1973: Shingle Street, Suffolk: an analysis of a geomorphic cycle. *Bulletin Geological Society of Norfolk*, 24, 15–35.

Ranwell, D. S. 1959: Newborough Warren, Angelesey. I. The dune system and dune slack habitat. *Journal of Ecology*, 47, 571–601.

Ranwell, D. S. 1964: *Spartina* salt marshes in southern England: II Rate and seasonal pattern of sediment accretion. *Journal of Ecology*, 52, 79–94.

Ranwell, D. S. 1972: *Ecology of salt marshes and sand dunes*. London: Chapman and Hall.

Ranwell, D. S. and Boar, R. 1986: *Coast Dune Management Guide*. Monks Wood: Institute of Terrestrial Ecology.

Redda, A., Hansom, J. D. and Brown, R. D. 1985: Edale End landslide. In D. J. Briggs, D. D. Gilbertson and R. D. S. Jenkinson, (eds), *Peak District and Northern Dukeries Field Guide*, Cambridge: Quaternary Research Association, 94–100.

Reid, C. 1884: Dust and soils. *Geological Magazine*, NS, 1, 165–169.

Reid, C. 1887: On the origin of the dry chalk valleys and of the coombe rock, *Quarterly Journal Geological Society of London*, 43, 364–73.

Reid, C. 1913: *Submerged Forests*. Cambridge, 129 pp.

Reid, E. M. and Chandler, M. E. J. 1933: *The London Clay Flora*. London: British Museum (Natural History).

Reynolds, B. 1986: A comparison of element outputs in solution, suspended sediments and bedload for a small upland catchment. *Earth Surface Processes and Landforms*, 11, 217–221.

Reynolds, C. S. 1979: The limnology of the eutrophic meres of the Shropshire-Cheshire plain: a review. *Field Studies*, 5, 93–173.

Richards, K. S. 1979: Channel adjustment to sediment pollution by the china clay industry in Cornwall, England. In D. D. Rhodes and G. P. Williams (eds), *Adjustments of the Fluvial System*, 309–31. Dubuque, Iowa: Kendall-Hunt.

Richards, K. S. 1981: Evidence of Flandrian valley alluviation in Staindale, N. Yorks Moors. *Earth Surface Processes and Landforms*, 6, 183–6.

Richards, K. S. and Anderson, M. G. 1978: Slope stability and valley formation in glacial outwash deposits, North Norfolk. *Earth Surface Processes*, 3, 301–18.

Richardson, J. A. 1976: Pit heap into pasture. In J. Lenihan and W. W. Fletcher (eds) *Reclamation*, 60–93. Glasgow: Blackie.

Richardson, S. J. and Smith, J. 1977: Peat wastage in the East Anglian Fens. *Journal of Soil Science*, 28, 485–9.

Riley, J. M. 1987: Drumlins of the southern Vale of Eden, Cumbria, England. In J. Menzies and J. Rose (eds), *Drumlin Symposium*, Rotterdam: Balkema, 323–33.

Roberts, M. C. 1985: The geomorphology and stratigraphy of the Lizard Loess in South Cornwall, England. *Boreas*, 14 (1), 75–82.

Robinson, A. H. W. 1955: The harbour entrances of Poole, Christchurch and Pagham. *Geographical Journal*, 121, 33–50.

Robinson, D. A. and Jerwood, L. C. 1987: Sub-aerial weathering of chalk shore platforms during harsh winters in Southeast England. *Marine Geology*, 77, 1–14.

Robinson, D. A. and Williams, R. B. G. 1976: Aspects of the geomorphology of the sandstone cliffs of the Central Weald. *Proceedings Geologists' Association of London*, 87, 93–9.

Robinson, D. A. and Williams, R. B. G. 1983: The Sussex coast past and present. In Geography Editorial Committee (ed.) *Sussex-Environment, Landscape and Society*. Gloucester: Alan Sutton, 50–66.

Robinson, D. A. and Williams, R. B. G. 1984: *Classic Landforms of the Weald* Geographical Association: Sheffield.

Robinson, D. N. 1968: Soil erosion by wind in Lincolnshire, March, 1968. *East Midland Geographer*, 4, 351–62.

Robinson, L. A. 1977: Marine erosive processes at the cliff foot. *Marine Geology*, 23, 257–71.

Robinson, L. A. 1977a: The morphology and development of the northeast Yorkshire shore platform. *Marine Geology*, 23, 237–55.

Robinson, L. A. 1977b: Erosive processes on the shore platform of northeast Yorkshire, England. *Marine Geology*, 23, 339–61.

Robinson, M. 1979: The effects of pre-afforestation ditching upon the water and sediment yields of a small upland catchment. *Working Paper, School of Geography, University of Leeds*, 252.

Robinson, M. 1986: Changes in catchment runoff following drainage and afforestation. *Journal of Hydrology*, 86, 71–84.

Robinson, M. and Blyth, K. 1982: The effect of forestry drainage operations on upland sediment yields: a case study. *Earth Surface Processes and Landforms*, 7, 85–90.

Robinson, M. A. and Lambrick, G. H. 1984: Holocene alluviation and hydrology in the Upper Thames Basin. *Nature*, 308, 809–14.

Rodda, J. C. 1970: Rainfall excesses in the United Kingdom. *Transactions, Institute of British Geographers*, 49, 49–60.

Rose, J. 1985: The Dimlington Stadial/Dimlington Chronozone: a proposal for naming the main glacial episode of the Late Devensian in Britain. *Boreas* 14(3), 225–30.

Rose, J. 1986: Weekend excursion to Southern East Anglia. *Mercian Geologist*, 10(2), 135–41.

Rose, J. 1987: Status of the Wolstonian Glaciation in the British Quaternary. *Quaternary Newsletter*, 53, 1–9.

Rose, J. and Allen, P. 1977: Middle Pleistocene stratigraphy in south-east Suffolk. *Journal Geological Society of London*, 133, 83–102.

Rose, L. and Vincent, P. 1986: The Kamenitzas of Gait Barrows National Nature Reserve, north Lancashire, England. In K. Paterson and M. M. Sweeting (eds), *New Directions in Karst*. Norwich: Geobooks, 473–96.

Rouse, W. C. and Bridges, E. M. 1985: Landslide susceptibility in the South Wales coalfield. In C. S. Morgan (ed.), *Landslides in the South Wales Coalfield*, Pontypridd: Polytechnic of Wales, 189–200.

Rouse, W. C. and Farhan, Y. I. 1976: Threshold slopes in South Wales. *Quarterly Journal of Engineering Geology*, 9, 327–38.

Rowe, P. 1988: Rates of incision in central England during the Quaternary. *Quaternary Newsletter*, 56, 21–32.

Roy, P. S. 1967: The recent sedimentology of Scolt Head Island, Norfolk, unpublished PhD thesis, University of London.

Rozier, I. T. and Reeves, M. J. 1979: Ground movement at Runswick Bay, north Yorkshire. *Earth Surface Processes*, 4, 275–80.

Ruddiman, W. F. and McIntyre, A. 1981: The North Atlantic ocean during the last deglaciation. *Palaeogeography, Palaeoclimatology, Palaeoecology*, 35, 145–214.

Savigear, R. A. G. 1952: Some observations on slope development in South Wales. *Transactions, Institute of British Geographers*, 18, 31–52.

Savigear, R. A. G. 1962: Some observations on slope development in north Devon and north Cornwall. *Transactions, Institute of British Geographers*, 31, 23–42.

Savin, S. M. 1977: The history of the Earth's surface temperatures during the past 100 million years. *Annual Review of Earth and Planetary Science* 5, 319–55.

Scaife, R. G. and Burrin, P. J. 1987: Further evidence for the environmental impact of prehistoric cultures in Sussex from alluvial fill deposits in the eastern Rother Valley. *Sussex Archaeological Collections*, 125, 1–9.

Sclater, J. G. and Christie, P. A. F. 1980: Continential stretching: an explanation of the post-mid-Cretaceous subsidence of the central North Sea basin. *Journal of Geophysical Research* 85, B, 3711–39.

Scourse, J. D. 1987: Periglacial sediments and landforms in the Isles of Scilly and west Cornwall. In J. Boardman (ed.), *Periglacial Processes and Landforms in Britain and Ireland*, Cambridge: Cambridge University Press, 225–36.

Scrope, G. P. 1866: The terraces of the Chalk Downs. *Geological Magazine*, *III*, 293–6.

Seddon, B. 1957: Late-glacial Cwm glaciers in Wales. *Journal of Glaciology* 3, 94–99.

Seddon, M. B. and Holyoak, D. T. 1985: Evidence of sustained regional permafrost during deposition of fossiliferous Late Pleistocene river sediments at Stanton Harcourt (Oxfordshire, England). *Proceedings Geologists' Association*, 96, 53–71.

Sejrup, H. P. and 9 co-workers. 1987: Quaternary stratigraphy of the Sladen area, central North Sea: a multi-disciplinary study. *Journal of Quaternary Science*, 2, 35–58.

Shackleton, N. J. and 16 others. 1984: Oxygen isotope calibration of the onset of ice-rafting and history of glaciation in the North Atlantic region. *Nature*, 307, 620–23.

Sharp, Q., Dowdeswell, J. A., and Gemmell, J. C. 1989: Reconstructing past glacier dynamics and erosion from glacial geomorphic evidence: Snowdon, North Wales. *Journal of Quaternary Science*, 4, 153–230.

Shaw, J. 1972: Sedimentation in the ice-contact environment from Shropshire (England). *Sedimentology*, 18, 23–62.

Shennan, I. 1982: Interpretation of Flandrian sea-level data from the Fenland, England. *Proceedings Geologists' Association*, 93, 53–63.

Shennan, I. 1983: Flandrian and late Devensian sea-level changes and crustal movements in England and Wales. In D. E. Smith and A. G. Dawson (eds), *Shorelines and Isostasy*, London: Academic Press, 255–83.

Shennan, I. 1986: Flandrian sea-level changes in the Fenland I: the geographical setting and evidence of relative sea-level changes. *Journal of Quaternary Science*, 1 (2), 119–53.

Shennan, I. 1986: Flandrian sea-level changes in the Fenland II: tendencies of sea-level movement, altitudinal changes, and local and regional factors. *Journal of Quaternary Science*, 1, 155–79.

Shephard-Thorn, E. R. 1975: The Quaternary of the Weald – a review. *Proceedings Geologists' Association of London*, 86, 537–47.

Shephard-Thorn, E. R. and Wymer, J. J. (eds) 1977: *INQUA Guidebook 1977 – SE England and N. and E. Thames Valley.*

Sherlock, R. L. 1931: *Man's Influence on the Earth.* London: Thornton Butterworth, 256 pp.

Shimwell, D. W. 1985: The distribution and origins of the lowland mosslands. In R. H. Johnson (ed.), *The geomorphology of north-west England.* Manchester. Manchester University Press, 298–312.

Shotton, F. W. 1953: The Pleistocene deposits of the area between Coventry, Rugby and Leamington and their bearing on the topographic development of the Midlands. *Philosophical Transactions of the Royal Society of London* B, 237, 209–60.

Shotton, F. W. 1960: Large scale patterned ground in the valley of the Worcestershire Avon. *Geological Magazine*, 97, 404–8.

Shotton, F. W. 1977: The English Midlands. *INQUA Excursion Guide A2, 10th INQUA Congress, Birmingham*, 51 pp.

Shotton, F. W. 1983: The Wolstonian stage of the British Pleistocene in and around its type area of the English Midlands. *Quaternary Science Reviews*, 2, 261–280.

Shotton, F. W. and Wilcockson, W. H. 1950: Superficial valley folds in an opencast working of the Barnsley Coal. *Proceedings Yorkshire Geological Society*, 28, 102–11.

Sissons, J. B. 1960a: Erosion surfaces, cyclic slopes and drainage systems in southern Scotland and northern England. *Transactions, Institute of British Geographers*, 28, 23–38.

Sisson, J. B. 1960/61: Some aspects of glacial drainage channels in Britain. *Scottish Geographical Magazine*, 76, 131–46; 77, 15–36.

Sissons, J. B. 1980: The Loch Lomond Advance in the Lake District, northern England. *Transactions, Royal Society of Edinburgh, Earth Sciences*, 71, 13–27.

Sissons, J. B. 1981: British shore platforms and ice sheets. *Nature*, 291, 473–5.

Skempton, A. W. and Weeks, A. G. 1976: The Quaternary history of the

Lower Greensand escarpment and Weald Clay vale near Sevenoaks, Kent. *Philosophical Transactions Royal Society of London* A, 283, 493–526.

Slaymaker, H. O. 1972: Patterns of present sub-aerial erosion and landforms in mid-Wales. *Transactions, Institute of British Geographers*, 55, 47–68.

Small, R. J. 1964: Geomorphology. In F. J. Monkhouse (ed.), *A Survey of Southampton and its Region*, Southampton: Southampton University Press, 37–50.

Small, R. J. 1965: The role of spring sapping in the formation of Chalk escarpment valleys. *Southampton Research Series in Geography*, 1, 3–29.

Small, R. J. 1970: *The Study of Landforms*. Cambridge: Cambridge University Press, 486 pp.

Small, R. J. 1980: The tertiary geomorphological evolution of south-east England: an alternative interpretation. *Institute of British Geographers Special Publication*, 11, 49–70.

Small, R. J., Clark, M. J. and Lewin, J. 1970: A periglacial rock-stream at Clatford Bottom, Marlborough Downs, Wiltshire. *Proceedings Geologists' Association*, 81, 87–98.

Small, R. J. and Fisher G. C. 1970: The origin of the secondary escarpment of the South Downs. *Transactions, Institute of British Geographers*, 49, 97–107.

Smart, J. G. O., Bissom, G. and Worssam, B. C. 1966: *The Geology of the Country around Canterbury and Folkestone*. Memoir, Geological Survey of UK.

Smart, P. L., Atkinson, T. C., Laidlaw, I. M. S., Newson, M. D. and Trudgill, S. T. 1986: Comparison of the results of quantitative and non-quantitative tracer tests for determination of karst conduit networks: an example from the Traligill Basin, Scotland. *Earth Surface Processes and Landforms*, 11, 249–261.

Smith, A. J. 1985: A catastrophic origin for the palaeovalley system of the eastern English Channel. *Marine Geology*, 64, 65–75.

Smith, B. J. and McAlister, J. J. 1987: Tertiary weathering environments and products in northeast Ireland. In V. Gardiner (ed.), *International Geomorphology 1986*, Part II, 1007–1031.

Smith, D. I. 1975a: The geomorphology of Mendip – the sculpting of the landscape. In D. I. Smith and D. P. Drew (eds) *Limestone and Caves of the Mendip Hills*. Newton Abbott: David and Charles, 89–132.

Smith, D. I. 1975b: The erosion of limestones on Mendip. In D. I. Smith and D. P. Drew (eds), *Limestones and Caves of the Mendip Hills*. Newton Abbott: David and Charles, pp. 130–47.

Smith, D. I. 1975c: The problem of limestone dry valleys – implications of recent work in limestone hydrology. In R. Peel, M. Chisholm, and P. Haggett (eds), *Processes in Physical and Human Geography*, 130–147.

Smith D. I. and Atkinson, T. C. 1976. Process, landform and climate in limestone regions. In E. Derbyshire (ed.) *Geomorphology and Climate*. Chichester: John Wiley, 367–409.

So, C. L. 1965: Coastal platforms of the Isle of Thanet, Kent. *Transactions, Institute of British Geographers*, 37, 147–56.

So, C. L. 1974: Some coast changes around Aberystwyth and Tanybwlch, Wales. *Transactions, Institute of British Geographers*, 62, 143–53.

Sparks, B. W. 1949: The denudation chronology of the dip-slope of the South Downs. *Proceedings Geologists' Association*, 60, 165–215.

Sparks, B. W. 1971: *Rocks and relief.* London: Longman, 404 pp.

Sparks, B. W. and West, R. G. 1964: The drift landforms around Holt, Norfolk. *Transactions, Institute of British Geographers*, 35, 27–35.

Sparks, B. W. and West, R. G. 1972: *The Ice-age in Britain.* London: Methuen.

Sparks, B. W., Williams, R. B. G. and Bell, F. G. 1972: Presumed ground-ice depressions in East Anglia. *Proceedings Royal Society of London*, A, 327, 329–43.

Sperling, C. H. B., Goudie, A. S., Stoddart, D. R. and Poole, G. G. 1977: Dolines of the Dorset Chalklands and other areas in Southern Britain. *Transactions, Institute of British Geographers*, NS2, 205–23.

Stamp. L. D. 1927: The Thames drainage system and the age of the Strait of Dover. *Geographical Journal*, 70, 386–90.

Starkel, L. 1966: The palaeogeography of mid- and eastern Europe during the last cold stage and west European comparisons. *Philosophical Transactions Royal Society of London* B, 280, 351–72.

Statham, I. 1976: Debris flows on vegetated screes in the Black Mountain, Carmarthenshire. *Earth Surface Processes and Landforms*, 1, 173–80.

Steers, J. A. 1948a: *The Coastline of England and Wales.* Cambridge: Cambridge University Press, 644 pp.

Steers, J. A. 1948b: Twelve years' measurement of accretion on Norfolk salt marshes. *Geological Magazine*, 85, 163–6.

Steers, J. A. 1960: *The Coast of England and Wales in Pictures.* Cambridge: Cambridge University Press, 146 pp.

Steers, J. A. 1977: Physiography. In V. J. Chapman (ed.), *Wet Coastal Ecosystems*, Amsterdam: Elsevier.

Steers, J. A. 1981: *Coastal features of England and Wales.* Cambridge: Oleander Press, 206 pp.

Steers, J. A. and Mitchell J. B. 1967: East Anglia. In J. B. Mitchell (ed.), *Great Britain: Geographical Essays.* Cambridge: Cambridge University Press pp. 86–103.

Stevenson, C. H. 1968: An analysis of the chemical composition of rain-water and air over the British Isles and Eire for the years 1959–1964. *Quarterly Journal of the Royal Meteorological Society*, 94, 56–70.

Straw, A. 1961: Drifts, meltwater channels and ice-margins in the Lincolnshire Wolds. *Transactions, Institute of British Geographers*, 29, 115–28.

Straw, A. 1970: Wind-gaps and water-gaps in Eastern England. *East Midlands Geographer*, 5, 97–106.

Stuart, A. 1924: The petrology of the dune sands of South Wales. *Proceedings Geologists' Association of London*, 35, 316–31.

Sugden, D. E. and John, B. S. 1976: *Glaciers and Landscape*. London: Arnold.

Sumbler, M. G. 1983: A new look at the type Wolstonian glacial deposits of Central England. *Proceedings Geologists' Association*, 94, 23–31.

Summerfield, M. A. and Goudie, A. S. 1980: The sarsens of southern England: their palaeoenvironmental interpretation with reference to other silcretes. *Institute of British Geographers Special Publication*, 11, 71–100.

Sutcliffe, D. W., Carrick, T. R., Heron, J., Rigg, E., Talling, J. F., Woof, C. and Lund, J. W. G. 1982: Long-term and seasonal changes in the chemical composition of precipitation and surface waters of lakes and tarns in the English Lake District. *Freshwater Biology*, 12, 451–506.

Sutherland, D. G. 1984: The Quaternary deposits and landforms of Scotland and the neighbouring shelves: a review. *Quaternary Science Reviews*, 3, 157–254.

Sweeting, M. M. 1950: Erosion cycles and limestone caverns in the Ingleborough District. *Geographical Journal*, 115, 63–78.

Sweeting, M. M. 1966: The weathering of limestones, with particular reference to the Carboniferous Limestones of northern England. In G. H. Dury (ed.), *Essays in Geomorphology*, London: Heinemann, 117–210.

Sweeting, M. M. 1970: Recent developments and techniques in the study of Karst landforms in the British Isles. *Geographia Polonica*, 8, 227–41.

Sweeting, M. M. 1972: Karst of Great Britain. In M. Herak and V. T. Stringfield (eds), *Karst: Important Karst Regions of the Northern Hemisphere*, Amsterdam: Elsevier, 417–43.

Sweeting, M. M. and Paterson, K. J. (eds) 1985: *New Directions in Karst*. Norwich: Geobooks.

Tallis, J. H. 1965: Studies on Southern Pennine peats, IV: Evidence of recent erosion. *Journal of Ecology*, 53, 509–20.

Tallis, J. H. 1973: Studies on Southern Pennine peats. V: Direct observation on peat erosion and peat hydrology at Featherbed Moss, Derbyshire. *Journal of Ecology*, 6, 1–22.

Tallis, J. H. 1985a: Mass movement and erosion of a southern Pennine blanket peat. *Journal of Ecology*, 73, 283–315.

Tallis, J. H. 1985b: Erosion of blanket peat in the southern Pennines: new light on an old problem. R. H. Johnson (ed.), *The Geomorphology of North-west England*, Manchester: Manchester University Press, 313–36

Taylor, B. J., Burgess, I. C., Land, D. H., Mills, D. A. C., Smith, D. B. and Warren, P. T. 1971: *Northern England*. London: Institute of Geological Sciences.

Taylor, C. 1975: *Fields in the English Landscape*. London: Dent, 174 pp.

Temple, P. H. 1965: Some aspects of cirque distribution in the west central Lake District, northern England. *Geografiska Annaler*, 47A, 185–93.

Te Punga, M. T. 1956: Altiplanation terraces in southern England. *Biulteyn Peryglacjalny*, 4, 331–8.

Te Punga, M. T. 1957: Periglaciation in southern England. *Tijdschrift Koninklijk Nederlandsch Aardrijkskundig Genootschap*, 74, 401–12.

Thomas, G. S. P. 1989: The Late Devensian glaciation along the western margin of the Cheshire–Shropshire lowland. *Journal of Quaternary Science*, 4, 167–81.

Thomas, T. M. 1956: Gully erosion in the Brecon Beacons area, South Wales. *Geography*, 41, 99–107.

Thomas, T. M. 1963: Solution subsidence in southeast Carmarthenshire and southwest Breconshire. *Transactions, Institute of British Geographers*, 33, 45–60.

Thomas, T. M. 1970a: The limestone pavements of the North Crop of the South Wales coalfield. *Transactions, Institute of British Geographers*, 50, 87–105.

Thomas, T. M. 1970b: The imprint of structural grain on the micro-relief of the Welsh upland. *Geographical Journal*, 7(1), 69–100.

Thomas, T. M. 1974: The South Wales interstratal karst. *Transactions, British Cave Research Association*, 1(3).

Thompson, D. B. and Worsley, P. 1967: Periods of ventifact formation in the Permo-Triassic and Quaternary of the North East Cheshire basin. *Mercian Geologist*, 2, 279–98.

Tinkler, K. J. 1966: Slope profiles and scree in the Eglwyseg Valley, North Wales. *Geographical Journal*, 132, 379–385.

Tinkler, K. J. 1985: *A Short History of Geomorphology*. London: Croom Helm, 317 pp.

Tomlinson, M. E. 1941: Pleistocene gravels on the Cotswolds Sub-edge Plain from Mickleton to the Frome Valley. *Quarterly Journal Geological Society*, 96, 385.

Tooley, M. J. 1978: *Sea-level Changes in North-west England During the Flandrian Stage*. Oxford: Clarendon Press, 232 pp.

Trechmann, C. T. 1920: On a deposit of interglacial loess, and some transported preglacial freshwater clays on the Durham coast. *Quarterly Journal Geological Society of London*, 75, 173–201.

Trenhaile, A. S. 1971: Lithological control of high-water rock ledges in the Vale of Glamorgan, Wales. *Geografiska Annaler*, 53A, 59–69.

Trenhaile, A. S. 1974a: The geometry of shore platforms in England and Wales. *Transactions, Institute of British Geographers*, 62, 129–42.

Trenhaile, A. S. 1974b: The morphology and classification of shore platforms in England and Wales. *Geografiska Annaler*, 56A, 103–10.

Trenhaile, A. S. 1987: *The Geomorphology of Rock Coasts*. Oxford: Clarendon Press.

Triccas, P. D. 1973: Periods of ventifact formation in the mid-Severn Valley. *Proceedings of the Birmingham Natural History Society*, 22, 199–207.

Trotter, F. M. 1929: The Tertiary uplift and resultant drainage of the

Alston Block and adjacent areas. *Proceedings of the Yorkshire Geological Society*, 21, 161–180.

Trudgill, S. 1985: *Limestone Geomorphology*. London: Longman.

Tubbs, C. 1984: Spartina on the south coast: an introduction. In P. Doody (ed.), *Spartina Anglica in Great Britain*, Shrewsbury: Nature Conservancy Council, 3–4.

Tuckfield, C. G. 1986: A study of dells in the New Forest, Hampshire, England. *Earth Surface Processes and Landforms*, 11, 23–40.

Tufnell, L. 1969: The range of periglacial phenomena in northern England. *Biulteyn Peryglacjalny*, 19, 291–323.

Tufnell, L. 1971: Erosion by snow particles in the north Pennines. *Weather*, 26, 492–8.

Tufnell, L. 1972: Ploughing blocks with special reference to north-west England. *Biulteyn Peryglacjalny*, 21, 237–70.

Tute, J. A. 1870: On certain natural pits in the neighbourhood of Ripon. *Report of the Proceedings of the Geological and Polytechnic Society of the West-Riding of Yorkshire*, 1868, 1–7.

Unwin, D. J. 1973: The distribution and orientation of corries in northern Snowdonia, Wales. *Transactions, Institute of British Geographers*, 58, 85–97.

Vincent, P. J. 1985: Quaternary geomorphology of the southern Lake District and Morecambe Bay area. In R. H. Johnson (ed.), *The Geomorphology of North-West England*, Manchester: Manchester University Press, 159–77.

Vincent, P. J. and Clarke, J. V. 1976: The terracette enigma – a review. *Biulteyn Peryglacjalny*, 25, 65–77.

Vincent, P. J. and Lee, M. P. 1981: Some observations on the loess around Morecambe Bay, North-west England. *Proceedings Yorkshire Geological Society*, 43(3), 281–94.

Vincent, P. J. and Lee, M. P. 1982: Snow patches on Farleton Fell, South-East Cumbria. *Geographical Journal*, 148, 337–42.

Walcott, R. I. 1972: Past sea levels, eustasy and the deformation of the Earth. *Quaternary Research*, 2, 1–14.

Walker, M. J. C. 1980: Late-glacial history of the Brecon Beacons, South Wales. *Nature*, 287, 133–5.

Walker, M. J. C. 1982: The Late Glacial and early Flandrian deposits at Traeth Mawr, Brecon Beacons, South Wales. *New Phytologist*, 90, 177–94.

Walling, D. E. 1971: Sediment dynamics of small instrumented catchments in South-east Devon. *Transactions Devon Association*, 103, 147–65.

Walling, D. E. 1979: The hydrological impact of building activity: a study near Exeter. In G. E. Hollis (ed.) *Man's impact on the hydrological cycle in the United Kingdom*, Norwich: Geo Abstracts, 135–51.

Walling, D. E. and Webb, B. W. 1981: Water Quality. In S. Lewin (ed.) *British Rivers*, London: Allen & Unwin, 126–69.

Walling, D. E. and Webb, B. W. 1986: Solutes in river systems. In S. T. Trudgill (ed.), *Solute Processes*, Chichester: John Wiley, 251–327.

Wallwork, K. L. 1956: Subsidence in the mid-Cheshire industrial area. *Geographical Journal*, 122, 40–53.

Wallwork, K. L. 1960: Some problems of subsidence and land use in the mid-Cheshire industrial area. *Geographical Journal*, 126, 191–9.

Wallwork, K. L. 1974: *Derelict Land*. Newton Abbott: David and Charles.

Walsh, P. T., Atkinson, K., Boulter, M. C. and Shakesby, R. A. 1987: The Oligocene and Miocene outliers of West Cornwall and their bearing on the geomorphological evolution of Oldland Britain. *Philosophical Transactions of the Royal Society of London* A, 323, 211–45.

Walsh, P. T., Boulter, M. C., Istaba, M. and Urbani, D. M. 1972: The preservation of the Neogene Brassington Formation of the southern Pennines and its bearing on the evolution of upland Britain. *Quarterly Journal Geological Society of London*, 128, 519–59.

Walsh, P. T. and Brown, E. H. 1971: Solution subsidence outliers containing probable Tertiary sediment in north-east Wales. *Geographical Journal*, 7, 299–320.

Walsh, R. P., Hudson, R. N. and Howells, K. A. 1982: Changes in the magnitude – frequency of flooding and heavy rainfalls in the Swansea Valley since 1875. *Cambria*, 9, 36–60.

Waltham, A. C. 1986: Valley Excavation in the Yorkshire Dales Karst. In K. Paterson and M. M. Sweeting (eds), *New Directions in Karst*, Norwich: Geobooks, 541–550.

Ward, J. C. 1870: On the denudation of the Lake District. *Geological Magazine*, Decade 1, 7, 14–17.

Ward, R. C. 1981: River systems and river regimes. In J. Lewin (ed.), *British Rivers*, London: Allen & Unwin, 1–33.

Ward, R. G. W. 1984: Avalanche prediction in Scotland: I. A survey of avalanche activity. *Applied Geography*, 4, 91–108.

Warren, P. T., Price, D., Nutt, M. J. C. and Smith, E. G. 1984: *Geology of the Country around Rhyl and Denbigh*. London: British Geological Survey.

Warwick, G. T. 1963: The dry valleys of the southern Pennines, England. *Erdkunde*, 18, 116–22.

Warwick, G. T. 1964: Relief and structure. In J. W. Watson and J. B. Sissons (eds), *The British Isles*, London: Nelson, 91–109.

Waters, R. S. 1960: The bearing of superficial deposits on the age and origin of the upland plain of East Devon, West Dorset and South Somerset. *Transactions, Institute of British Geographers*, 28, 89–97.

Waters, R. S. 1962: Altiplanation terraces and slope development in west Spitsbergen and south-west England. *Biulteyn Peryglacjalny*, 11, 89–101.

Waters, R. S. 1978: Periglacial geomorphology in Britain. In C. Embleton, D. Brunsden and D. K. C. Jones (eds), *Geomorphology: Present Problems and Future Prospects*, Oxford: Oxford University Press, 154–61.

Waton, P. V. and Barber, K. E. 1987: Rimsmoor, Dorset: biostratigraphy and chronology of an infilled doline. In K. E. Barber (ed.), *Wessex and the Isle of Wight Field Guide*, Cambridge: QRA, 75–80.

Watson, A. 1976: The origin and distribution of closed depressions in south-west Lancashire and north-west Cheshire, unpublished BA dissertation, University of Oxford.

Watson, E. 1960: Glacial landforms in the Cader Idris area. *Geography*, 45, 27–38.

Watson, E. 1977: The periglacial environment of Great Britain during the Devensian. *Philosophical Transactions Royal Society of London*, B, 280, 183–98.

Watson, E. 1981: Characteristics of ice-wedge casts in west central Wales. *Biulteyn Peryglacjalny*, 28, 163–77.

Watson, E. and Watson, S. 1967: The periglacial origin of the drifts at Morfa Bychan near Aberystwth. *Geological Journal*, 5, 419–40.

Watson, E. and Watson, S. 1974. Remains of pingos in the Cletwr Basin, Southwest Wales. *Geografiska Annaler*, 56A, 213–25.

Watson, E. and Watson. S. 1977: *INQUA Congress Guidebook for Excursion 69. Mid and North Wales.*

Watts, W. W. 1903: A buried Triassic landscape. *Geographical Journal*, 21, 623–36.

Watts, W. W. 1910: Charnwood Forest. In H. W. Monckton and R. Herries (eds), *Geology in the Field*, London: Stanford, 770–85.

Waylen, M. J. 1979: Chemical weathering in a drainage basin underlain by Old Red Sandstone. *Earth Surface Processes*, 4(2), 167–78.

Weeks, A. G. 1969: The stability of natural slopes in south-east England as affected by periglacial activity. *Quarterly Journal Engineering Geology*, 2, 49–61.

Wells, S. G. and Harvey, A. M. 1987: Sedimentologic and geomorphic variations in storm generated alluvial fans, Howgill Fells, northwest England. *Bulletin Geological Society of America*, 98, 182–98.

West, G. and Dumbleton, M. J. 1972: Some observations on swallow holes and mines in the chalk. *Quarterly Journal Engineering Geology*, 8 (1, 2), 171–8.

Weyman, D. R. 1974: Runoff process, contributing area and streamflow in a small upland catchment. *Institute of British Geographers Special Publication*, 6, 33–43.

Wheeler, D. A. 1979: The overall shape of longitudinal profiles of streams. In A. F. Pitty (ed.), *Geographical Approaches to Fluvial Processes*, Norwich: Geo Abstracts, 241–260.

Whitaker, W. 1921: *The Water Supply of Norfolk from Underground Sources.* London: Memoir, Geological Survey of UK.

Whiteman, A., Naylor, D., Pegrum, R. and Rees, G. 1975: North Sea Troughs and Plate Tectonics. *Tectonophysics*, 26, 39–54.

Whittow, J. B. and Hardy, J. R. 1971: Revised edition of A. E. Trueman's *Geology and Scenery in England and Wales*. Harmondsworth: Penguin Books.

Williams, A. G., Ternan, L. and Kent, M. 1986: Some observations on the chemical weathering of the Dartmoor granite. *Earth Surface Processes and Landforms*, 11, 557–574.

Williams, G. and Hall, M. 1987: The loss of coastal grazing marshes in south and east England with special reference to East Essex, England. *Biological Conservation*, 39, 243–54.

Williams, G. J. 1968: The buried channel and superficial deposits of the Lower Usk, and their correlations with similar features in the Lower Severn. *Proceedings Geologists' Association*, 79, 325–48.

Williams, P. W. 1966: Limestone pavements with special reference to Western Ireland. *Transactions, Institute of British Geographers*, 40, 155–72.

Williams, R. B. G. 1964: Fossil patterned ground in eastern England. *Biulteyn Peryglacjalny*, 14, 337–49.

Williams, R. B. G. 1968: Some estimates of periglacial erosion in southern and eastern England. *Biulteyn Peryglacjalny*, 17, 311–35.

Williams, R. B. G. 1969: Permafrost and temperature conditions in England during the last glacial period. In T. L. Péwé (ed.), *The Periglacial Environment*, Montreal: McGill-Queen's University Press, 339–410.

Williams, R. B. G. 1975: The British climate during the last glaciation: an interpretation based on periglacial phenomena. In A. E. Wright and F. Moseley (eds), *Ice Ages: Ancient and Modern*, Liverpool: Seel House Press, 95–120.

Williams, R. B. G. 1980: The weathering and erosion of chalk under periglacial conditions. In D. K. C. Jones (ed.), *The Shaping of Southern England*, London: Academic Press, 225–48.

Williams, R. B. G. and Robinson, D. A. 1983: The landforms of Sussex. In *The Geography Editorial Committee Sussex: Environment, Landscape and Society*, Gloucester: Alan Sutton, 33–49.

Williams, R. B. G. 1986: Periglacial phenomena in the South Downs. In G. de G. Sieveking and M. B. Hart (eds), *The Scientific Study of Flint and Chalk*, Cambridge: Cambridge University Press, 161–7.

Wills, L. J. 1924: The development of the Severn Valley in the neighbourhood of Iron-Bridge and Bridgnorth. *Quarterly Journal Geological Society of London*, 80, 274–314.

Wills, L. J. 1938: The Pleistocene development of the Severn from Bridgnorth to the sea. *Quarterly Journal of the Geological Society of London*, 94, 161–242.

Wilson, K. 1960: The time factor in the development of dune soils at South Haven peninsula, Dorset. *Journal of Ecology*, 48, 341–59.

Wilson, P., Bateman, R. M. and Catt, J. A. 1981: Petrography, origin and environment of deposition of the Shirdley Hill Sand of southwest Lancashire, England. *Proceedings Geologists' Association*, 92, 211–29.

Winslow, J. H. 1966: Raised submarine canyons: an exploratory hypothesis. *Annals of the Association of American Geographers*, 56, 634–72.

Wintle, A. G. 1981: Thermoluminescence dating of late Devensian loesses in southern England. *Nature*, 289, 479–80.

Wolfe, J. A. 1978: A palaeobotanical interpretation of Tertiary climates in the Northern Hemisphere. *American Scientist*, 66, 694–703.

Wood, A. 1959: The erosional history of the cliffs around Aberystywth. *Liverpool and Manchester Geological Journal*, 2, 271–87.

Wood, A. 1968: Beach platforms in the Chalk of Kent, England. *Zeitschrift für Geomorphologie*, 12, 107–13.

Wood, E. S. 1972: *Field Guide to Archaeology*, 3rd edn. London: Collins, 384 pp.

Wood, S. V. 1882: The newer Pliocene period in England. *Quarterly Journal Geological Society of London*, 38, 667–745.

Woodland, A. W. 1970: The buried tunnel-valleys of East Anglia. *Proceedings Yorkshire Geological Society*, 37, 521–78.

Wooldridge, S. W. 1938: The glaciation of the London Basin and the evolution of the Lower Thames drainage system. *Quarterly Journal Geological Society of London*, 94, 627–67.

Wooldridge, S. W. 1950: The upland plains of Britain; their origin and geographic significance. *Advancement of Science*, 7, 162–75.

Wooldridge, S. W. 1952: The changing physical landscape of Britain. *Geographical Journal*, 118, 297–308.

Wooldridge, S. W. 1954: The physique of the South West. *Geography*, 39, 231–242.

Wooldridge, S. W. 1958: The trend of geomorphology. *Transactions, Institute of British Geographers*, 25, 29–35.

Wooldridge, S. W. and Linton, D. L. 1939: Structure, surface and drainage in South-east England. *Transactions, Institute of British Geographers*, 10, i–xvi, 1–124.

Wooldridge, S. W. and Linton, D. L. 1955: *Structure, Surface and Drainage in South-east England*. London: George Philip, 176 pp.

Worsley, P. 1966: Fossil frost wedge polygons at Congleton, Cheshire, England. *Geografiska Annaler*, 48A, 211–19.

Worsley, P. 1975: An appraisal of the glacial Lake Lapworth concept. In A. D. M. Phillips and B. J. Turton (eds.), *Environment, Man and Economic Change*, London: Longman, 98–118.

Worsley, P. 1977: Periglaciation. In F. W. Shotton (ed.), *British Quaternary Studies: Recent Advances*, Oxford: Clarendon Press, 205–19.

Worssam, B. C. 1963: *Geology of the Country around Maidstone*. Memoir of the Geological Survey of Great Britain. London: HMSO.

Wright, L. W. 1969: Shore platforms and mass movement: a note. *Earth Science Journal*, 3, 44–50.

Wright, L. W. 1970: Variation in the level of the cliff/shore platform junction along the south coast of Great Britain. *Marine Geology*, 9, 347–53.

Wright, V. P. 1986: The polyphase Karstification of the Carboniferous Limestone in South Wales. In K. Paterson and M. M. Sweeting (eds), *New Directions in Karst*, Norwich: Geobooks, 569–80.

Wright, W. B. 1937: *The Quaternary Ice Age*. London: Macmillan, 478 pp.

Wyatt, R. J., Horton, A. and Kenna, R. J. 1971: Drift-filled channels on the Leicestershire-Lincolnshire border. *Bulletin Geological Survey of Great Britain*, 37, 57–79.

Yates, E. M. 1956: The Keele surface and the Upper Trent Drainage. *East Midland Geographer*, 1(5), 10–22.

Yates, E. M. and Moseley, F. 1958: Glacial lakes and spillways in the vicinity of Medeley, north Staffordshire. *Quarterly Journal Geological Society of London*, 113, 409–28.

Young, A. 1960: Soil movement by denudational processes on slopes. *Nature*, 188, 120–2.

Young, A. 1961: Characteristic and limiting slope angles. *Zeitschrift für Geomorphologie*, 5, 126–31.

Young, A. 1974: The role of slope retreat. *Institute of British Geographers Special Publication*, 7, 65–78.

Young, A. 1978: A twelve-year record of soil movement on a slope. *Zeitschrift für Geomorphologie Supplementband*, 29, 104–10.

Ziegler, P. A. 1978: North-western Europe: tectonics and basin development. *Geologie en Mijnbouw*, 57, 589–626.

INDEX

Note: References to names of major landforms are limited to those discussed in detail in the text and/or in diagrams, tables or illustrations. Others are subsumed under entries for major regions and types of landforms.